Tamilselvi Kaliyamoorthi
Mangaleeswaran R.
Bharath MR.

Access of water, Sanitation and Hygiene (WASH) in Urban areas in Tamil

Tamilselvi Kaliyamoorthi
Mangaleeswaran R.
Bharath MR.

Access of water, Sanitation and Hygiene (WASH) in Urban areas in Tamil

RUSA Base-line Survey of WASH in Perambalur, Ariyalur & Nilgiris in Tamil Nadu

LAP LAMBERT Academic Publishing

Imprint
Any brand names and product names mentioned in this book are subject to trademark, brand or patent protection and are trademarks or registered trademarks of their respective holders. The use of brand names, product names, common names, trade names, product descriptions etc. even without a particular marking in this work is in no way to be construed to mean that such names may be regarded as unrestricted in respect of trademark and brand protection legislation and could thus be used by anyone.

Cover image: www.ingimage.com

Publisher:
LAP LAMBERT Academic Publishing
is a trademark of
Dodo Books Indian Ocean Ltd. and OmniScriptum S.R.L publishing group

120 High Road, East Finchley, London, N2 9ED, United Kingdom
Str. Armeneasca 28/1, office 1, Chisinau MD-2012, Republic of Moldova, Europe
Managing Directors: Ieva Konstantinova, Victoria Ursu
info@omniscriptum.com

Printed at: see last page
ISBN: 978-620-8-44074-9

I Dedicated The Book to My Beloved 'Amma' who inspired me to go to work

Dr.Salai Ponsusheela,
Farmer Superintend of, Edn.Dept.TN

I am not fortunate enough to be in her womb, but I thank God for giving me the fortune to live in her heart. Is there a better word in the world than the word "Amma"? Mother is a better word than the word "God." My mother, who kept me safe in the sinful age of Kali Yuga and provided me with a way to reach God, has nothing to offer you. Mother, I offer you, my tears.

Contents

Foreword 5

Preface...7

Acknowledgments 11

Prologue 13

1.Introduction 15

2. Demographic and Socio-Economic Characteristics .19

3.Access of Water.. 29

4. Sanitation.. 41

5. Environmental & Personal Hygiene......................... 63

Findings & Conclusion.. 88

Foreword

Improving Water, Sanitation, and Hygiene (WASH) and infrastructure conditions is an important component of Sustainability in India. The RUSA 2.0 acknowledges that for the effective intervention in this regard it is necessary to make a system assessment of the existing situation, promote the distinct and detailed standards of Sanitary-Hygiene and water conditions in urban areas like Municipality and Town Panchayat. This is a study about Access of water, Sanitation and Hygiene in District of Tamil Nadu. The RUSA 2.0 Conducted study in Nilgiris district's urban areas. The baseline survey was conducted in Municipalities and Town panchayats, the total sample size is 4552. The survey included access of water, sanitation and hygiene. The study provides a general situational overview of availability of water, access of toilet facilities, and hygienic practice of the people.

The demographic and socioeconomic characteristics of Ariyalur, Perambalur, and Nilgiris districts reveal diverse residential patterns, gender distributions, and family structures. Ariyalur predominantly consists of municipalities, Perambalur has a slight majority of town panchayats, and Nilgiris shows an equal split between town panchayats and municipalities. Gender distribution varies, with Nilgiris having a higher male population, Perambalur more females, and Ariyalur relatively balanced. Nuclear families dominate all districts, with Nilgiris and Ariyalur having the highest percentages.

Socioeconomic data indicates that Ariyalur and Perambalur have a Hindu majority, while Nilgiris has a notable Christian population. Occupational distributions show a prevalence of manual labor in Ariyalur and Perambalur, with skilled labor more common in Nilgiris. Income disparities exist, with Nilgiris having a higher proportion in the middle-income bracket compared to Ariyalur and Perambalur.

This book of Dr.K. Tamilselvi is an important contribution in this field of research. This research work is based on a well designed covering a sample of about 1500 household belonging to urban areas of three districts – Ariyalur, Perambalur and Nilgiris – in Tamil Nadu. Some findings are worthy to note and some others area astounding I strongly believe that the readers of this book will get some ideas of the process of access of water, sanitation and hygiene (WASH) in urban areas of Tail Nadu. Overall, this is good document for academicians and reaserchers engaged in teaching and research on WASH.

Dr. Nirmala Saravanan

Assistant Professor,

Indira Gandhi Medical College & Research Institute,

Puducherry.

01/02/2025

Preface

Water access and quality vary significantly across the districts. Public taps are the main source in Nilgiris, surface water in Ariyalur, and borewells in Perambalur. Water treatment practices, primarily boiling, are common but inconsistent, with many households not treating their water, posing health risks. The quality of water varies, with issues like sediment, colored water, and poor taste or odor reported, particularly in Perambalur and Ariyalur.

Sanitation facilities have improved, with a high percentage of households having access to toilets, mostly Indian-style. However, challenges in maintenance and gender disparities in sanitation responsibilities persist. Open defecation remains an issue, particularly in Ariyalur. Personal hygiene practices show good adherence to certain behaviors, like food preparation hygiene and handwashing, but indicate areas needing improvement, especially in bathing frequency and menstrual hygiene management. To address these challenges, targeted interventions in infrastructure development, community education, and sustainable practices are essential to enhance overall well-being and equity among residents.

Ariyalur predominantly consists of Municipalities (68.8%), while Perambalur has a slight majority of Town Panchayats (52.5%). Nilgiris shows an equal split between Town Panchayats and Municipalities (50% each). Nilgiris has a higher male population (54.8%), Perambalur has more females (56.7%), and Ariyalur shows a balanced gender distribution. The majority of the population in all districts falls within the 25-54 age group. Nuclear families are predominant across all districts, Perambalur has a higher proportion of joint families (19.0%). Ariyalur and Perambalur have a Hindu majority, while Nilgiris has a diverse composition with a notable Christian population. Perambalur also has a higher percentage of Muslims. Backward Classes (BC) dominate all three districts, with Scheduled Castes (SC) forming a significant portion in Nilgiris.

Ariyalur shows a prevalence of skilled manual labor, while unskilled manual labor is more common in Perambalur and Nilgiris. Housewives constitute a notable demographic in Ariyalur. A substantial proportion of the population in all districts falls within the lowest income bracket (<=1000). Concrete houses are predominant in all districts, indicating relatively good living conditions. However, Nilgiris has a notable proportion of asbestos/steel houses (22.3%). Home ownership is high across all districts, with Ariyalur leading at 81.3%.

Rented houses are more common in Perambalur and Nilgiris. Almost all households in Ariyalur own their homestead land (99.8%), while Perambalur and Nilgiris rely more on government-owned land. Public taps are the dominant source for drinking and non-drinking purposes in Nilgiris (68.7%), whereas Ariyalur relies more on surface water sources like ponds and streams (90.9%). Perambalur shows a balanced reliance on public taps (18.5%) and borewells (54.3%). Individual household taps are prevalent in all districts. Ariyalur significantly uses surface water (90.9%), while Perambalur (64.7%) and Nilgiris (30.1%) show varied preferences, influenced by local geography and infrastructure.

Ariyalur and Perambalur experience limited daily water supply (1-5 hours/day) more frequently than Nilgiris. Females predominantly handle water fetching responsibilities, particularly in Ariyalur and Perambalur, highlighting gender disparities in water management. Boiling is the most common method across all districts (38.9%), Significant proportions (23.5%) do not treat their water, posing potential health risks. Nilgiris exhibits higher boiling practices (54.3%), while Ariyalur reports the highest non-treatment rate (47.6%). Ariyalur shows higher sediment presence (14.6%) and better taste/odor perception (50.4%). Perambalur reports more colored (6.6%) and salty water (7.8%), while Nilgiris has the highest percentage of colorless water (75.1%) and the lowest chlorinated water presence (0.7%). With 83.0% of households in Nilgiris having access to toilets at home. Indian-style toilets are predominant, most toilets were constructed within the past decade, indicating ongoing sanitation infrastructure development. Septic tanks are widely used, though maintenance practices vary. Significant percentages in all districts never clean their tanks, highlighting potential hygiene risks and the need for improved maintenance education and services. While most respondents reported no major issues with their toilets, a minority faced challenges such as septic tank overflow, broken basins, and pipe blockages.

Despite progress, open defecation remains a concern, particularly in Ariyalur (19.4%), Perambalur (12.5%), and Nilgiris (12.3%) districts. This study on environmental hygiene and waste management across Ariyalur, Nilgiris, and Perambalur districts reveals critical insights into solid and liquid waste practices, highlighting both successes and challenges in environmental sustainability efforts. Hygienic practices during food handling are widely practiced. Hand washing practices are generally well-adhered to, particularly before eating and after using the toilet, although there is room for improvement in consistent hand hygiene across different daily activities. Personal hygiene practices like bathing and teeth brushing show cultural and regional variations. Access to private bathing facilities is widespread, although some households still rely on open spaces for bathing, indicating disparities in sanitation infrastructure. During menstruation, bathing habits vary, reflecting cultural norms and personal preferences, with some women maintaining regular bathing routines while others adhere to specific practices during their menstrual period. There is a strong call for improved facilities in educational institutions and workplaces, including access to sanitary vending machines, designated changing spaces, and adequate water supply. Awareness programs are also recommended to address misconceptions and promote better menstrual health practices. Headers in the front matter of the book contain the name of the chapter alone.

Dr.K.Tamilselvi

01/03/2025

Acknowledgments

At the outset I express my sincere thanks deep gratitude to and dedicated to my beloved Mother **Dr.Salai Ponsusheela,** Superintend, Education Department Tamil Nadu (Rtd), Meyvazhisalai and I am dedicating this book to her who inspired me to go to work. I am extremely thankful to grandfather **Meyvazhi Ilam Kalai Kottu Andandar and my Brother Salai Shanmugasundaram, BSNL,** Meyvazhisalai for their moral supporting.

I am thankful to Professor Dr.R.Manimegalai, Professor and Head, Department of Women Studies, to given the golden opportunity to work as Research Associate in the RUSA 2.0 project. I am also thankful to **Prof. Dr.R. Mangaleswaran,** Professor and head, Department of Social Work, **Prof. Dr. K. Suriyans,** Professor and Head, Department of Sociology, for his continuous encouragement and support to finish this research work and complete the book work. I express my sincere gratitude to My Ph.D guide **Prof.Dr.S.Ramachandaran ,** Department Population studies, Annamalai University, **Prof. Dr.N. Audinarayana,** Department Population studies, Bharathiyar University, Coimbatore , who taught me the intricacies of the study and helped me by checking my analysis tables.

I am thankful to the M.R. Bharath, Project Fellow, Central Theme RUSA 2.0 to supporting to complete the work, Ms.Rama Prabha Project Fellow, Sociology Department And I thankful to Dr.Malaiyandi, Research Associate, Women Studies RUSA 2.0 and Dr.Sathish Kumar, Project Fellow, Centre for Youth Development and Studies.

My special thanks to my friends K.Selvam, Dr.Vani Archana, Dr. Nirmala. Parents, daughter and thankful to RUSA 2.0 SS, Bharathidasan University.

Prologue

A Prologue, in case of fiction books, can be an excerpt from the book or a section of the story that happens before the true beginning of the story. It could provide readers an introduction to the nature of the characters in the story, a view of events that happened earlier and also set the tone of the book.

An Introduction, in case of non-fiction books, defines what the book is about, what a reader can expect from the book, the motivation for the book and, sometimes, guidance on how to use the book.

As you can see here, the first paragraph is usually given a flush left alignment (Normal_without indent) and all subsequent paragraphs are given a slight indent (Normal_indent). The first paragraph of a new section after a subheading should also be given a flush left alignment. This rule applies to all chapters in the book.

Headers in the front matter of the book contain the name of the chapter alone. Pages in which the chapter starts would not have headers or footers.

You can delete this page if you do not wish to have a Prologue or Introduction in the book.

We've added additional text in this section to show you how the headers and footers appear in the subsequent pages. Please delete the instructions and the additional text when you add the actual content of the book.

Lorem ipsum dolor sit amet, consectetur adipiscing elit. Sed ornare tristique mauris at varius. Phasellus arcu leo, tristique quis nulla id, ultrices varius diam. Morbi vulputate lacinia odio, sed tristique est sollicitudin a. Nullam fringilla dui sit amet pretium facilisis. Ut in neque dui. Aliquam vehicula sem id gravida finibus. Pellentesque sodales efficitur arcu, at dapibus enim porta vitae. Duis porttitor euismod eros elementum auctor. Praesent libero tortor, gravida ac fringilla sed, lobortis a dui. Nullam bibendum condimentum urna. Duis tincidunt imperdiet ipsum, vitae feugiat urna. Nulla facilisi. Curabitur ut ultricies justo, ac pretium sem. Aliquam aliquet ipsum quis tortor ultrices, vitae posuere orci porta. Curabitur dignissim risus vitae felis lobortis vehicula. In tempus tincidunt diam, vel malesuada leo fringilla nec.

Fusce laoreet ipsum sem, a vestibulum turpis vestibulum et. Donec venenatis finibus tincidunt. Quisque placerat tellus lectus, eu consequat tellus suscipit vitae. Curabitur pharetra libero diam, at imperdiet dolor malesuada congue. Sed ac consectetur elit. Suspendisse eu faucibus quam. Mauris volutpat molestie urna. Duis vestibulum ante vel odio efficitur, vehicula euismod ex malesuada.

1. Introduction

Water availability per person is dependent on population of the country and for India, per capita water availability in the country is reducing due to increase in population. The average annual per capita water availability in the years 2001 and 2011 was assessed as 1816 cubic meters and 1545 cubic meters respectively which may further reduce to 1486 cubic meters and 1367 cubic meters in the years 2021 and 2031 respectively (Ministry of Jal Shakti, 2020). Indians using 80 per cent of good water used for agriculture, 5 per cent water using for domestic purpose and 15 per cent of the water using for factories (NITI Aayog, 2018).

India currently stores only 6% of its annual rainfall or 253 billion cubic metres (8.9×1012 cu ft), while developed nations strategically store 250% of the annual rainfall in arid river basins. India also relies excessively on groundwater resources, which accounts for over 50 percent of irrigated area with 20 million tube wells installed. India has built nearly 5,000 major or medium dams, barrages, etc. to store the river waters and enhance ground water recharging. The important dams (59 nos) have an aggregate gross storage capacity of 170 billion cubic metres (6.0×1012 cu ft). About 15 percent of India's food is being produced using rapidly depleting /mining groundwater resources. The end of the era of massive expansion in groundwater use is going to demand greater reliance on surface water supply systems.

As per Ministry of Housing and Urban Affairs, 135 litres per capita per day (lpcd) has been suggested as the benchmark for urban water supply. For rural areas, a minimum service delivery of 55 lpcd has been fixed under Jal Jeevan Mission, which may be enhanced to higher level by states. Indian Council for Agriculture Research (ICAR) Director General T Mohapatra pointed out that the per capita annual water availability has declined to 1,508 cubic meter in 2014 from 5,177 cubic meter in 1951. The per capita availability of water is estimated to decline further to 1,465 cubic meters by 2025 and 1,235 cubic meter by 2050. If it declines further to around 1,000-1,100 cubic meters, then India could be declared as water-stressed country. (The Hindu Business Line, 2019). Because of Indian' using the ground water then other sources. But they lack of storage habit would lead to the water issues.

Objective of the Study

To conduct a KAP and Baseline Need assessment Survey on WASH, employment and skill training needs in the sample districts.

To analyses the socio-economic profile of participants and identify the WASH needs of urban and peri urban habitats and the gender specific WASH needs of Women, adolescent girls, elderly, persons with disability, transgender persons.

To assess access to work place sanitation needs including gender specific needs and disposal of waste facilities in the formal and informal sector, the gender challenges and related loss of economic participation say construction work, street and road side vendors, road laying site workers, urban poor settlements, homeless, platform dwellers and other vulnerable groups.

To find out the soft and hard components of sanitation structure to ensure collective behavioral change, engaging men through Gender Sensitization, Interpersonal Communication and Psychological trauma, safety, security, dignity of persons and provide an enabling environment t to break and articulate on the culture of silence on gender specific needs.

To identify the gap on the SGD 5 and SDG 6 indicators and targets on WASH and Gender as defined in the National Framework and Baseline Survey on SDGs and assess the gap if any and plan for working towards the

To examine the implementation of sanitation facilities as per the legal provisions of Acts or laws and assess the gap if any in the work and public places in a gender perspective and advocate for implementation and of the legal provisions if there is any gap.

To develop WASH, Gender and Economic Participation Indices based on the indicators modified by the Government of Tamil Nadu.

To offer skill development and sensitization on the knowledge of managing sanitary structures among the local volunteers and ensure enabling environment for governance.

To document, publish and disseminate the findings through research papers, working papers, occasional papers, books, etc. Suggest Strategies for co-operatively managed rain water harvesting, collective and volunteer engagement of civil society on improving the sewage, sanitation and solid waste management, joint income generating activities, skill development, training on maintenance of local facilities, etc. and document the best practices.

Methodology

The Study will be conducted in Tamil Nadu, confining to Tamil Nadu. Going by the Human Development Index, there are four bottom districts as per HDR 2017. As per the Agro Climatic Zone division, there are seven zones. The sample districts will be selected as per the agro climatic zone wise, and also considering the low HDI performing district. Accordingly, seven districts, say Nagapatinam (Cavery Delta), Perambalur (North Eastern), Theni (Western), Karur (North Western), Virudhunagar (Southern), Nilgiris (Hilly Area) and Kanyakumari (High Rainfall Area) will be selected as sample districts. In addition, Ariyalur, which is in the low rank of HDI will be covered totally constituting ten districts. 2 percent of the households will be selected from the total number of households block wise covering both urban and peri-urban areas.

Study area Profile: The Nilgiris District, known as "the queen of hills" is situated at an elevation of 900-2636meters above Mean Sea Level (MSL). Approximately 65% of geographical area of this district lies at an altitude of 1800-2500 meters above MSL and the remaining 35% at lower altitudes. The climate is temperate to sub-tropical. The annual rainfall ranges from 950-1550mms.

Above study was conducted in Ooty Municipality and Ketty Town Panchayat. Those areas select for compare study research problems. Udhagamandalam, popularly known as Ooty, it is the capital of Nilgiris District. It is nestled at an altitude of 2240 meters above Mean Sea Level and spread out at 11.24 degrees North latitude and 76.44-degree longitude. Ketti is a selection Grade Town Panchayat, constituted in the year 1920. The Town is coming under the administrative territory of Nilgiris District. Because 75 per cent of Agricultural land is surround in this area. The Ketti Town Panchayat situated at the top of View point of Udhagamandalam. The Ketti Town Panchayat has 64 habitations, 18 wards and 120 streets. The total extent is 19.20 sqkm with total population of 762141 out of which male 360143 and female 375251 as per 2011 census.

Sample selection:

The multi-stage sampling procedure was adopted to choose the respondents for the study purpose. It means that the sample selection was made at different stages: selection state, district, and Municipalities and Town-panchayat in Nilgiris district. The respondents both gender in the age from 18 and above by the residence. First, house listing was conducted (a preliminary survey) to identify the houses. The first case was selected from which every "n" th case (systematic number) was selected till achieving the desired sample size form each area like Municipalities and Town Panchayat.

Instruments and Data collection

Method and Tools of Data collection were decided to administer survey schedule as a tool of data collection through personal interview with the respondents by using the Electronic Tablets for Data collection for the survey. An interview schedule was prepared in line with the objectives. All the questions are structed; close ended and multi responses questions related to the following aspects: i) Socio-economic background, ii) Availability of water, iii) Sanitation: Toilets facilities, iv) Hygiene: Public hygiene and personal hygiene.

2. Demographic and Socio-Economic Characteristics

We can experience the of Five Elements power but we cannot see the Five Elements Power

\- *Dr.K.Tamilselvi*

You can enter the text of the second chapter here.

The provided Table 1's data represents the distribution of residential areas by two type like Town Panchayat and Municipality, across three districts: Nilgiris, Perambalur, and Ariyalur. (Table 1). In Nilgiris, the distribution between Town Panchayat and Municipality is equal, each constituting 50% of the residential areas. In contrast, Perambalur has a slight majority of Town Panchayat areas (52.5%) compared to Municipalities (47.5%). Ariyalur shows a different pattern, with a significantly higher proportion of Municipalities (68.8%) compared to Town Panchayat areas (31.2%).

Table 2.1

District wise Percentage distribution of study areas

Types of Residential area	District		
	Ariyalur	Perambalur	Nilgiris
Town Panchayat	342 (31.2)	645 (52.5)	1112 (50.0)
Municipality	754 (68.8)	584(47.5)	1114 (50.0)

The Table 2.2 shows the demographic characteristics of the respondent like gender, age, number of household members, marital status and educational status. Nilgiris has a higher male population (54.8%) compared to females (45.2%), whereas Perambalur has more females

(56.7%) than males (42.9%). Ariyalur has a fairly balanced gender distribution with a slight male majority (52.3%). The majority of the population in all three districts falls within the 25-54 age group. Nilgiris has a relatively higher proportion of elderly (65+) at 16.2%, compared to 15.2% in Perambalur and 12.9% in Ariyalur. Nuclear families are predominant across all districts, especially in Nilgiris (93.5%) and Ariyalur (91.2%). Perambalur has a higher proportion of joint families (19.0%) compared to the other districts. Most households consist of 3-4 persons in all districts. Nilgiris has the highest proportion of smaller households (1-2 persons) at 40.4%, whereas larger households (5+ persons) are least common in Nilgiris (12.4%). A large majority of the population is married in all three districts. The highest proportion of widows is found in Nilgiris (11.8%). Education levels vary significantly. Nilgiris has the highest percentage of individuals with primary education (21.3%), middle school education (12.6%), and SSLC (23.9%). Perambalur has the highest percentage of individuals with higher secondary education (10.9%), while Ariyalur has a relatively higher proportion of postgraduates (11.9%).

A person or group of related or unrelated persons who live together in the same dwelling unit(s), who acknowledge one adult male or female as the head of the household, who share the same housekeeping arrangements, and who are considered a single unit. As per the De facto population all persons who stayed in the selected households the night before the interview (whether usual residents or visitors). De jure population All persons who are usual residents of the selected households, whether or not they stayed in the household the night before the interview. This data is based on the de facto population. Table 2.2 shows that on the average number of households 3.4. The average household size decreased slightly between 2005-06 and 2015-16 from 4.8 versus 4.6 persons. (NFHS-4) Nearly thirty percent of the households has 2 persons in a house. Rare families had above the 4 to 9 persons in a household.

Table 2.2

Percentage distribution of Demographic Characteristics in District Wise

Demographic Characteristics		Ariyalur		Perambalur		Nilgiris	
		N	%	N	%	N	%
Gender	Male	574	52.3	527	42.9	1219	54.8
	Female	523	47.7	697	56.7	1007	45.2
	Transgender	-	-	5	0.4	-	-
Age	18-19	7	0.6	12	1.0	17	0.8
	20-24	45	4.2	49	4.0	76	3.4
	25-54	654	60.6	729	59.3	1305	58.6
	55-64	235	21.8	252	20.5	468	21.0
	65+	139	12.9	187	15.2	360	16.2
Nature of family	Nuclear	1000	91.2	995	81.0	2082	93.5
	Joint	144	6.5	234	19.0	97	8.8
Household size	1-2 Persons	302	27.5	408	33.2	900	40.4
	3-4 Persons	590	53.8	583	47.4	1050	47.2
	5+ Persons	205	18.7	238	19.4	276	12.4
Marital Status	Married	892	81.3	1028	83.6	1784	80.1
	Unmarried	88	8.0	69	5.6	141	6.3
	Separated	8	0.7	16	1.3	33	1.5
	Widow	108	9.8	116	9.4	263	11.8

	Divorced	1	0.1	-	-	5	0.2
Education	Illiterate	209	19.1	175	14.2	269	12.1
	Primary	211	19.2	295	24.0	475	21.3
	Middle	116	10.6	145	11.8	281	12.6
	SSLC	292	26.6	249	20.3	532	23.9
	HSS	-	-	134	10.9	319	14.3
	UG	122	11.1	135	11.0	252	11.3
	PG	130	11.9	96	7.8	98	4.4

2.2 Socio-Economic characteristics:

According to Durkhiem "religion as a unified system of beliefs and practices related to sacred things, that is to say, things get apart and forbidden". (Bhusan vidhya and sachedeva D.R, 1984). The demographic characteristics of Ariyalur, Perambalur, and Nilgiris districts reveal significant insights into the socioeconomic status, religious composition, and occupational distribution within these regions.

Table 2.3

Percentage Distribution of Socio-Economic characteristics in District wise

Socio-Economic characteristics		**Ariyalur**		**Perambalur**		**Nilgiris**	
		N	%	N	%	N	%
Religion	Hindu	870	79.3	937	76.2	1629	73.2
	Christian	183	16.7	22	1.8	448	20.1

	Muslim	149	6.7	270	22.0	44	4.0
Community	BC	573	52.2	689	56.1	1244	55.9
	OC	3	0.3	84	6.8	28	1.3
	MBC	348	31.7	195	15.9	73	3.3
	SC	161	14.7	254	20.7	863	38.8
	ST	12	1.1	7	0.6	18	0.8
Occupation	Housewife	471	38.0	22	1.8	42	1.9
	Professional	17	1.5	136	11.1	126	5.7
	Clerical	3	0.3	81	6.6	241	10.8
	Sales	122	11.1	45	3.7	116	5.2
	Skilled Manual	538	49.0	202	16.4	547	24.6
	Unskilled Manual	-	-	743	60.5	1150	51.7
Income	<=1000	511	46.6	611	49.7	826	37.1
	1001-5000	158	14.4	166	13.5	319	14.3
	5001-10000	175	16.0	198	16.1	826	37.1
	10001-15000	114	10.4	70	5.7	241	10.8
	15001-20000	37	3.4	54	4.4	110	4.9
	20001-25000	33	3.0	21	1.7	33	1.5
	>25000	69	6.3	109	8.9	75	3.4

Ariyalur and Perambalur have a Hindu majority, while Nilgiris has a more diverse religious composition with a notable Christian population. Perambalur has a higher percentage of Muslims compared to the other districts. The Backward Class (BC) is the dominant community in all three districts. Scheduled Castes (SC) form a significant portion of the population in Nilgiris, much higher than in Ariyalur and Perambalur. Skilled manual labor is predominant in Ariyalur, whereas unskilled manual labor is more common in Perambalur and Nilgiris. There is a noticeable proportion of housewives in Ariyalur compared to the other districts. A significant portion of the population in all three districts falls within the lowest income bracket (<=1000), indicating a need for economic development and poverty alleviation programs. The income distribution shows that a considerable number of individuals in Nilgiris have a middle-range income (5001-10000), indicating a relatively better

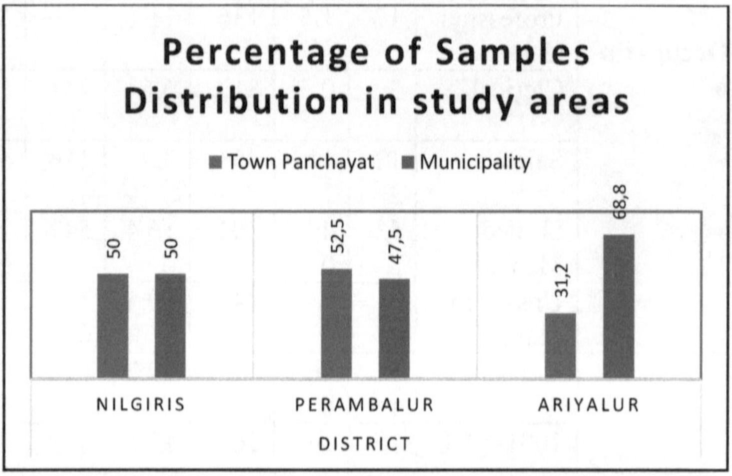

Fig. 1. Percentage of samples Distribution in study areas

economic status compared to the other districts. Previous studies have highlighted the correlation between demographic characteristics and socioeconomic development. For instance, Kumar and Singh (2018) found that regions with higher literacy rates and diverse occupational structures tend to have better economic growth.

The predominance of manual labor in Ariyalur and Perambalur suggests a need for skill development programs to transition the workforce towards more skilled professions. According to the National Sample Survey Office (NSSO) reports, regions with higher concentrations of backward and scheduled castes often face challenges in accessing quality education and employment opportunities. This is evident in the community distribution of Nilgiris, where a significant SC population may require targeted social and economic upliftment programs. Classification of occupation is used in the national context for the collection and dissemination of statistics from sources such as population censuses, labour market surveys, employer surveys, and other sources. It is also used by the Government and other organizations in activities such as administration of worker's compensation, matching jobseekers with job vacancies, and management of employment-related migration.

The classification of occupation is a categorization of individual occupations on the kind of work and skill-level involved in that particular occupation. It is intended that all the job titles in the nation can be assigned to one (and only one) of these categories or codes. In this study purposively categorized the occupation as professional/technical/administrative/managerial, clerical, sales & services, skilled manual, unskilled manual and agriculture and Housewife. Housewife means, "a woman who does not have a job outside the home and who spends her time cleaning the house, cooking, looking after her family", etc.

2.3 Housing details

Table 2.4

Percentage distribution of Housing Details in District Wise

Demographic Characteristics		**Ariyalur**		**Perambalur**		**Nilgiris**	
		N	%	N	%	N	%
Type of House	Thatched	61	5.6	65	5.3	81	3.6
	Tiled	337	30.7	195	15.9	613	27.5
	Concrete	648	56.7	648	59.1	1036	46.5
	Asbestos/steel	51	4.6	71	5.8	496	22.3

Ownership of house	Own	892	81.3	917	74.6	1605	74.1
	Rented house	201	18.3	294	23.9	476	21.4
	Leased house	3	0.3	-	-	24	1.1
	Govt. Scheme	-	-	18	1.5	72	3.2
	NGO	1	0.1	-	-	4	0.2
Ownership of homestead	Own land	1095	99.8	917	74.6	1650	74.1
	Govt. Land	2	0.2	294	23.9	476	21.4
	NGO- Land	-	-	-	-	24	1.1
	Religious Institution land	-	-	18	1.5	76	3.4

The type of housing and ownership patterns provide insights into the socioeconomic status and stability of households. According to studies by Desai and Jain (2017), owning a house and homestead land significantly contributes to the economic stability and social status of families. Concrete houses indicate better living conditions compared to thatched or asbestos/steel houses, which are associated with poorer households (World Bank, 2018). Furthermore, the reliance on government and NGO schemes for housing indicates the role of social safety nets in supporting vulnerable populations. Research by Sharma and Singh (2019) highlights the importance of such schemes in providing affordable housing to low-income families and reducing homelessness.

The majority of houses in Ariyalur, Perambalur, and Nilgiris are made of concrete, indicating relatively good housing conditions. However, a notable proportion of houses in Nilgiris are made of asbestos/steel (22.3%), which raises concerns about health and safety due to potential exposure to harmful materials. Most houses in the three districts are owned by the residents, with Ariyalur having the highest percentage of home ownership (81.3%). Rented houses are more common in Perambalur (23.9%) and Nilgiris (21.4%). Nearly all households in Ariyalur own their homestead land (99.8%), while a significant portion of households in Perambalur and Nilgiris rely on government land (23.9% and 21.4%, respectively).

3. Access of Water

3.1 Sources of Water drinking and non-drinking purpose

This chapter deals about the drinking and non-drinking water, and explaining sources, timing, days gaps between availability, protection of at home, collecting, quality and payment of the water. Access to safe drinking water is crucial for public health, economic development, and overall quality of life. According to WHO, a significant portion of the global population still relies on unimproved water sources, which are prone to contamination and associated health risks. Studies have shown that the type of water source can significantly influence the incidence of waterborne diseases.

Figure 3.1

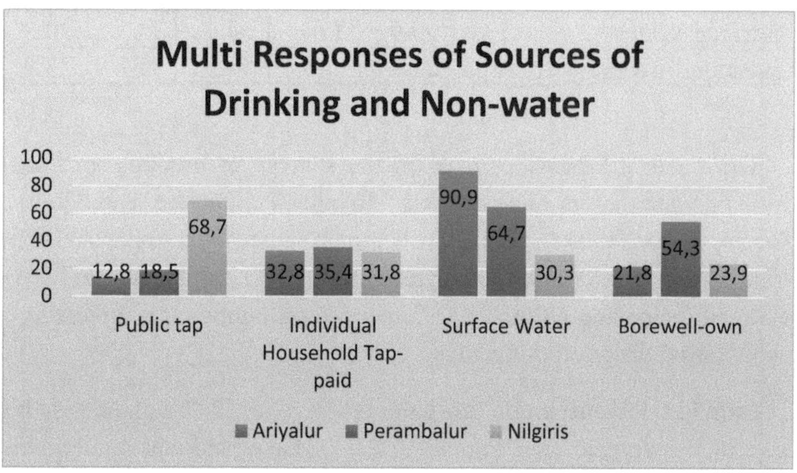

Study from other parts of the Tamil Nadu reported similar variation, nearly half the households get drinking water from a piped source in the hamlets (small villages) of the tribal area of Jawadhi hills. Raman Sivakumar (2000) studied on the physico-chemical analysis of water sources of Nilgiris, in this study pointed out the water sources of Ooty are many lakes, man-made impoundment and open wells. Recent developmental activities in and around Ooty had serious impact on the water quality of these surface water-bodies. In the present study the physico chemical analysis of water from few lakes, open wells and streams, for two seasons during 1997, was conducted. The parameters including total hardness, COD, Chlorides and heavy metals were also carried out. Comparisons were made with statutory standards and mitigatory measures are suggested for improving quality.

Table 3.1

Multi Responses of Sources of Drinking and Non-water

Sources of Drinking and Non-water	Ariyalur		Perambalur		Nilgiris	
	N	%	N	%	N	%
Public tap	207	12.8	300	18.5	1113	68.7
Individual Household Tap-paid	817	32.8	882	35.4	793	31.8
Surface Water	73	90.9	134	64.7	71	30.3
Borewell-own	138	21.8	343	54.3	151	23.9

The Table 3.1 provides data on the sources of drinking and non-drinking water across three districts: Ariyalur, Perambalur, and Nilgiris. The population relies on public taps for main sources of drinking and non-drinking purpose is **Public Tap** in **Ariyalur** 12.8%, in **Perambalur**18.5% use public taps and **Nilgiris** 68.7% depend on public taps, indicating a significant reliance on this source.

Individual Household Tap-paid in Ariyalur 32.8% of households have their own paid taps, Peramblur 35.4% have paid taps and Nilgiris 31.8% use this source, showing relatively balanced use among the three districts. The surface water sources of Surface like Pond, Lake, Stream and River water 90.9 % used by the Ariyalur district people, 64.7 per cent used by the Perambalur and Nilgiris 30.1 % was recorded, reflecting a different water source preference. 21.8% use personal borewells in Ariyalur, 54.3% rely on borewells, indicating significant usage and 23.9% use borewells, showing moderate reliance. In regions like Ariyalur, Perambalur, and Nilgiris, water source preferences reflect both availability and socio-economic factors. The high reliance on public taps in Nilgiris suggests either a well-maintained public infrastructure or limited access to alternative sources. In contrast, Perambalur's significant use of borewells and surface water indicates varying availability and perhaps better access to these sources compared to public taps.

Access to safe water is essential for health and socio-economic development. According to the WHO, disparities in water access can perpetuate health inequities. Social factors, including caste, often influence access to water sources in India, reflecting historical and socio-economic disparities. Studies indicate that marginalized groups, such as SCs and BCs, frequently rely on less secure water sources, increasing their risk of waterborne diseases.

Table 3.2

District wise – Cross classification of Percentage of Multi Responses of Sources of Water

Primary Sources of Water	Ariyalur				Perambalur			
	Community				Community			
	SC	MBC	BC	OC	SC	MBC	BC	OC
	%	%	%	%	%	%	%	%
Public Tap	9.4	41.8	11.5	-	16.1	35.5	15.8	75.8
Tap-Paid	17.0	61.8	30.8	41	28.4	29.4	37.7	75.7
Borewell	15.8	59.8	15.4	-	32.7	37.0	61.4	100.
Surface Water	114.8	131.0	94.7	-	6.5	58.6	64.0	87.5

Table 3.2 (a)

District wise – Cross classification of Percentage of Multi Responses of Sources of Water

Primary Sources of Water	**Nilgiris**			
	Community			
	SC	MBC	BC	OC
	%	%	%	%
Public Tap	74.5	22.7	72.7	24.2
Tap-Paid	54.5	8.8	31.5	20.3
Borewell	51.5	3.3	22.7	-
Surface Water	16.7	10.3	41.3	12.5

The Table 3.2 presents data on the primary sources of water across different social categories (Scheduled Castes - SC, Most Backward Classes - MBC, Backward Classes - BC, and Other Castes - OC) in three districts: Ariyalur, Perambalur, and Nilgiris. The percentages indicate the proportion of each social category relying on various water sources: public tap, paid tap, borewell, and surface water. The data reveals significant variations in water source dependency across different socialcategories and districts. MBC and BC groups in Ariyalur show high dependence on surface water and paid taps. Perambalur's OC group heavily relies on public taps, paid taps, and borewells, whereas Nilgiris displays a high reliance on public taps across all categories, especially SC and BC. Surface water usage is notable among MBC in Ariyalur and BC in Nilgiris. The high reliance on public taps among SCs in Nilgiris and OCs in Perambalur might indicate improved public infrastructure in these regions. However, the significant use of surface water among MBCs in Ariyalur suggests challenges in accessing safer water sources. The data underscores the need for targeted interventions to ensure equitable access to safe water across all social categories.

3.2 Access of Water:

Table 3.3

District wise Percentage distribution of access of water

Access of Water		Ariyalur		Perambalur		Nilgiris	
		N	%	N	%	N	%
Average Hours of Water supply per day	Always	318	29.0	171	13.9	341	15.3
	1-5 hours	756	68.9	1052	85.6	1121	50.4
	6-10 hours	13	1.2	2	0.2	190	8.5
	> 10 hours	10	0.9	4	0.3	574	25.8
Water supply in a week	Daily	280	25.5	181	14.7	1209	54.3
	2-5 days gap	465	42.4	426	34.7	757	34.0
	6-10 days gap	352	32.1	552	44.9	200	9.0
	>10 days gap	-	-	70	5.7	60	2.7
Water Fetching	Male	471	42.9	293	23.8	831	37.3
	Female	626	57.1	936	76.2	1395	62.7
Payment for drinking water	Yes	846	77.1	906	73.7	725	32.6
	No	198	18.0	323	26.3	1501	67.4

The table 3.3 presents data on the distribution of water access in three districts: Ariyalur, Perambalur, and Nilgiris. It covers average hours of water supply per day, frequency of water supply in a week, water fetching responsibilities by gender, and payment for drinking water. The data highlights significant variations in water access and management across Ariyalur, Perambalur, and Nilgiris. Ariyalur and Perambalur show high percentages of limited water supply (1-5 hours per day) and irregular weekly supply, with a notable gender disparity in water-fetching responsibilities, primarily falling on females. Nilgiris, however, has a better daily supply, though a substantial portion still experiences limited supply (1-5 hours per day). Payment for drinking water is common in Ariyalur and Perambalur, whereas the majority in Nilgiris do not pay for water.

Access to reliable water supply is crucial for health, sanitation, and economic activities. According to the World Health Organization (WHO), water scarcity and irregular supply significantly impact daily living and health outcomes. In many developing regions, women disproportionately bear the burden of water fetching, affecting their time for education and economic activities. Payment for water, while necessary for maintenance, can impose a financial burden on low-income households. Studies have shown that regular and sufficient water supply can improve health outcomes, reduce time spent on water collection, and increase productivity. Gender disparities in water collection often reflect broader socio-economic inequalities.

A previous studies indicate the optimum benefit from water supply in rural and urban areas can only achieve if communities and individuals are participated and change their attitudes by made aware of the links between hygiene practices, polluted water source, save water and disease (Muthachitra & B.Laxmi Kanta 2017). Times of India daily news publish the Nilgiris tribal village becomes model for clean drinking water, the Banagudi Kurumba village near Kotagiri in the Nilgiris has paved the way of the community to sustain successfully as 'Nanner gramam' it was the first tribal village in the state to get purified piped water using ultra-filtration membrane-based purification system (NAWA, 2021)

3.3 Water Treatment at home

Table 3.4

Multi Responses of Treatment of Drinking water at Home

Treatment of water at home[a]	N	%
Filter (cloth, strainer)	1679	22.9%
Boiling	2844	38.9%
Reverse osmosis (RO)	695	9.5%
Chlorinated	384	5.2%

No treatment	1717	23.5%
Total	7319	100.0%
a. Dichotomy group tabulated at value 1.		

Access to clean and safe drinking water is critical for preventing waterborne diseases and ensuring public health. According to the World Health Organization (WHO), water treatment at the household level is essential, especially in areas where centralized water treatment systems are inadequate or non-existent. Common household water treatment methods include boiling, filtration, chlorination, and advanced purification technologies like reverse osmosis (RO). Studies have shown that boiling water is highly effective in killing pathogens and is widely adopted in many regions due to its simplicity and effectiveness. Filtration methods, such as cloth or strainer filters, can remove particulate matter but are less effective against microorganisms. Reverse osmosis (RO) is a sophisticated method that effectively removes contaminants, including dissolved solids and microorganisms, but is less accessible due to higher costs. Chlorination is an effective method for disinfecting water but is less commonly used at the household level due to concerns about the taste and potential chemical residues. The practice of not treating water at home poses significant health risks, particularly in areas where water sources are contaminated. The prevalence of untreated water consumption highlights the need for improved awareness and accessibility to effective water treatment methods.

Figure 3.2

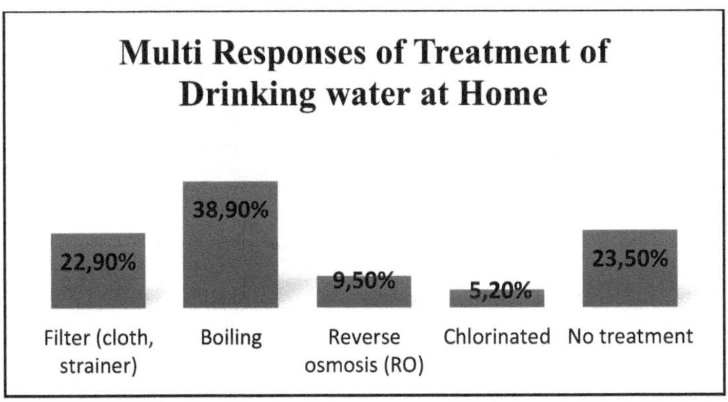

An experimental study was indicated that the most popular brands in Malaysia is Brand A from RO filtration system and the home treatment of tap water to ensure that the quality of water physically look better and safe to be consumed. (Nur Shahirah Abd Rahim & Norzila Othman, 2019).

The table 3.4 presents data on the various methods used for treating drinking water at home, based on multiple responses from a sample of 7319 households.

The percentages indicate the proportion of households using each method: Filter (cloth, strainer): 22.9%, Boiling: 38.9%, Reverse Osmosis (RO): 9.5%, Chlorinated: 5.2% and No treatment directly used: 23.5%. The data reveals that the most common method of treating drinking water at home is boiling, utilized by 38.9% of households. A significant portion of households (23.5%) do not treat their water, indicating potential exposure to waterborne contaminants. The use of cloth or strainer filters and reverse osmosis (RO) systems are also prevalent, while chlorination is the least common method.

Table 3.5

Cross classification of water treatment with community						
Multi response of water Treatment[a]	District					
	Ariyalur		Perambalur		Nilgiris	
	N	%	N	%	N	%
Filter (cloth, strainer)	330	24.6	230	11.7	1119	27.9
Boiling	320	23.9	344	17.5	2180	54.3
Reverse osmosis (RO)	48	3.6	418	21.3	229	5.7
Chlorinated	3	0.2	157	8.0	224	5.6
No treatment	638	47.6	818	41.6	261	6.5
Percentages and totals are based on responses.						
a. Dichotomy group tabulated at value 1.						

The table 3.5 presents data on the methods of drinking water treatment across three districts: Ariyalur, Perambalur, and Nilgiris. The percentages reflect the proportion of respondents within each district using various treatment methods: like filter, boiling, RO etc. The data reveals significant differences in water treatment practices across the three districts. Nilgiris has the highest percentage of households boiling water (54.3%) and the lowest percentage of households not treating their water (6.5%). In contrast, Ariyalur has the highest percentage of households not treating their water (47.6%) and a relatively low use of reverse osmosis systems (3.6%). Perambalur shows a higher use of reverse osmosis (21.3%) and chlorination (8.0%) compared to the other districts.

3.4 Characteristics of Water:

According to the guidelines "Water quality refers to the Chemical, physical, and biological and characteristics of water based on the standards of its usage" (WHO, 2022). In this study shows about the different characteristics of the water. Access to clean and safe drinking water is a critical public health issue. According to the World Health Organization (WHO), the quality of drinking water can significantly impact health outcomes. Characteristics such as color, transparency, taste, and odor are important indicators of water quality. Senthilnathan S (1999) study was point out the water quality of effluents in Tirupur, with respect to all these parameters, exceeded the tolerance limit for irrigation and public usage. Heavy metals in the effluent are in the order of Zn> Cr> Cu> Cd. Generally, water that is colorless and transparent is free from visible contaminants and is typically considered safe for consumption. The presence of colored water can indicate contamination from organic materials, metals, or other pollutants, which may pose health risks. Sediment in water can result from inadequate filtration or contamination from environmental sources, which can harbor harmful microorganisms and reduce water quality.

Acceptable taste and odor are important for consumer satisfaction and health. Unpleasant taste or odor can result from chemical contamination or biological growth in water sources. Chlorinated water indicates the presence of a disinfectant used to kill harmful pathogens. Low levels of chlorinated water usage may point to a lack of access to or acceptance of this treatment method. Salty water can be due to high concentrations of dissolved salts, often resulting from natural mineral deposits or pollution. Consumption of high-salinity water can lead to health issues. The table 3.6 presents the characteristics of drinking water based on responses from a sample of 5658 households.

Table 3.6

Multi Responses of the characteristic of Water

Characteristics of Water[a]	N	%
Colorless and transparent	3316	58.6%
Colored water	474	8.4%
Sediment	491	8.7%
Acceptable taste and odor less	1245	22.0%
Chlorinated water	32	0.6%
Salty	100	1.8%
Total	5658	100.0%

a. Dichotomy group tabulated at value 1.

Figure 3.3

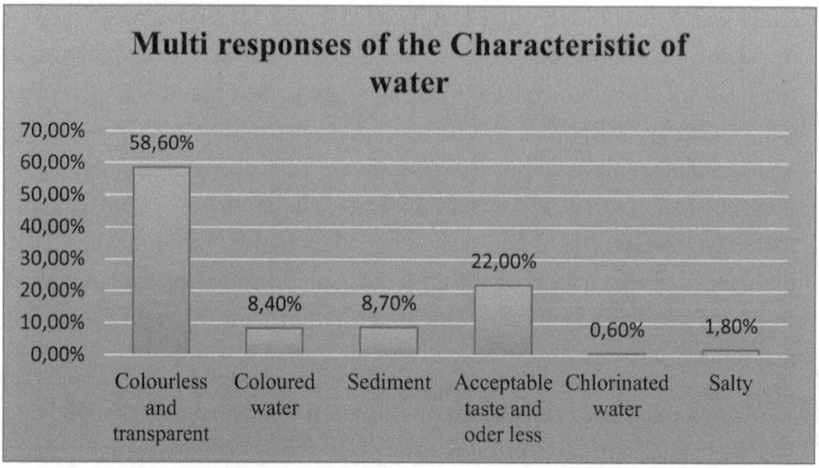

The percentages represent the proportion of households reporting each characteristic of their drinking water: The data indicates that the majority of households (58.6%) have access to colorless and transparent drinking water, which is generally considered a sign of good water quality. However, a significant portion of households report issues such as sediment (8.7%), colored water (8.4%), and unacceptable taste or odor (22.0%). The presence of chlorinated water is minimal (0.6%), suggesting limited use of chlorination as a treatment method. Additionally, a small percentage of households (1.8%) report salty water, which can indicate high mineral content or contamination.

Table 3.7

District Wise Multi Responses of Characteristics of Water

Multi response of Characteristics of water[a]	District					
	Ariyalur		Perambalur		Nilgiris	
	N	%	N	%	N	%
Colorless	658	32.1	908	71.2	1750	75.1
Colored water	59	2.9	84	6.6	331	14.2
Sediment	300	14.6	55	4.3	136	5.8
Acceptable taste	1034	50.4	114	8.9	97	4.2
Chlorinated water	-	-	15	1.2	17	0.7
Salty	-	-	100	7.8	-	-
Total	2051		1276		2331	

Percentages and totals are based on responses.

a. Dichotomy group tabulated at value 1.

Study from other parts of the country Study on drinking water quality at public utilities in Visakhapatnam, India, (2015, Sanjay Singh, R. et.al) study was conducted and tested the water in Lab and find out the sample areas water poor quality, likewise in this Nilgiris study not tested the water quality but the quality meant by the coloring of water and taste only, but according to the scientific test is must to find the quality of water. Access to clean and safe drinking water is crucial for public health. The quality of drinking water is influenced by various factors including geographical location, water sources, treatment methods, and infrastructure. Water quality indicators such as color, transparency, sediment, taste, odor, and chemical composition play essential roles in determining water safety. Clear and transparent water is typically associated with lower levels of suspended solids and microbial contaminants, indicating better water quality and safety. Colored water may indicate the presence of organic matter, metals, or other pollutants, potentially affecting both water safety and aesthetic qualities. Sediment in water can result from soil erosion, inadequate filtration, or pollution, potentially harboring harmful microorganisms and reducing water clarity. Acceptable taste and odor are critical for consumer acceptance and indicate the absence of chemical contaminants or biological growth that could affect water safety. Chlorination is a common method to disinfect water and kill pathogens. The low percentage of chlorinated water in the districts suggests varying levels of access to or adoption of water treatment practices. High salinity in water can result from natural mineral deposits, seawater intrusion, or industrial pollution, posing health risks and impacting water usability.

The table 3.7 presents the characteristics of drinking water across three districts: Ariyalur, Perambalur, and Nilgiris, based on multiple responses from households. The percentages represent the proportion of households reporting each characteristic of their drinking water: color less, colored, sediment, acceptable taste and odorless and chlorinated and salty water. The data reveals significant variations in the quality of drinking water across the three districts. Ariyalur generally reports higher percentages of sediment (14.6%) and a larger proportion of households with acceptable taste and odorless water (50.4%). Perambalur shows a higher incidence of colored water (6.6%) and salty water (7.8%), while Nilgiris reports the highest percentage of households with colorless and transparent water (75.1%) and the lowest presence of chlorinated water (0.7%).

4. Sanitation

4.1 Availability of Toilet at Home

This chapter deals about the availability, uses of toilets, issues and maintenance of toilet at home level. Access to sanitation facilities is a fundamental aspect of public health and human dignity. Improved sanitation contributes significantly to reducing waterborne diseases, improving hygiene practices, and enhancing overall quality of life. The United Nations Sustainable Development Goals (SDGs) emphasize universal access to adequate and equitable sanitation and hygiene by 2030, aiming to ensure availability and sustainable management of water and sanitation for all. Access to toilets not only reduces open defecation but also promotes health and hygiene, especially in rural and underserved areas. High coverage rates, as observed in the data, contribute to improved community health outcomes and environmental cleanliness. The choice between Indian and Western-style toilets often reflects cultural preferences, affordability, and infrastructure availability. While Indian-style toilets are more common in many regions due to cost-effectiveness and ease of maintenance, the adoption of Western-style toilets signifies rising standards of living and changing societal norms. Gender dynamics play a crucial role in sanitation practices, with women often bearing the burden of sanitation-related responsibilities. Involving both men and women in toilet construction initiatives promotes gender equality and ensures that sanitation solutions meet the needs of all household members.

The Table 4.1 provides a comprehensive overview of sanitation facilities across three districts: Ariyalur, Perambalur, and Nilgiris. It includes data on the availability, use, type of toilets, and the involvement of individuals in toilet construction: Availability of toilet, use of toilet, types of toilets, persons who insisted to build the toilet at home etc. The data highlights substantial progress in sanitation access across Ariyalur, Perambalur, and Nilgiris, with a high percentage of households having access to toilets. The use of toilets is also prevalent, indicating good utilization of available facilities. Indian-style toilets are predominant, though some households have adopted Western-style toilets or use both types. Involvement in toilet construction shows varied gender participation, with notable female involvement in Perambalur and Nilgiris. In Nilgiris district 83.0 per cent of people have toilet facility at home, a similar study was told around 69 per cent of the sample household has toilets facility in the home functional in rural areas Manjakuttai, in Salem district (Raja Sekaran N, 2019). A study on promotion on construction and uses was mentioned the per centage of improvement of the uses was increased, there was a significant reduction in perceived psychosocial barriers in toilet adoption (S Nancy et.al., 2022).

Table 4.1

District Wise Percentage distribution of Toilet Facility

Toilet Facility	Ariyalur		Perambalur		Nilgiris	
	N	%	N	%	N	%
Availability of Toilet						
Yes	848	77.3	1046	85.1	1882	84.5
No	249	22.7	183	14.9	344	15.5
Total	1097	100.0	1229	100.0	2226	100.0
Use of Toilet						
NA	249	22.7	-	-	-	-
Yes	845	77.0	1037	84.4	1875	84.2
No	3	0.3	192	15.6	351	15.8
Total	1097	100.0	1229	100.0	2226	100.0
Type of Toilet						
NA	249	22.7	183	14.9	344	15.5
Indian Type	702	64.0	799	65.0	1589	71.4
Western Type	69	6.3	101	8.2	158	7.1
Both	77	7.0	146	11.9	135	6.1
Total	1097	100.0	1229	100.0	2226	100.0

Persons insisted to build the toilet at home						
No Idea	183	16.7	-	-	-	-
Male	613	55.9	719	58.5	1229	55.5
Female	17	1.5	327	26.6	653	29.3
NA	284	25.9	183	14.9	344	15.5
Total	1097	100.0	1229	100.0	2226	100.0

Access to adequate sanitation is crucial for human health, dignity, and sustainable development. Improved sanitation reduces the spread of diseases, particularly waterborne illnesses, and enhances overall community well-being. The World Health Organization (WHO) emphasizes the importance of safe sanitation facilities and proper waste management to achieve global health goals. Financing options and subsidies play a crucial role in promoting toilet construction, particularly in low-income and marginalized communities.

Government initiatives and NGO partnerships are instrumental in supporting household investments in sanitation. Properly maintained septic tanks are essential for safely managing human waste and preventing environmental contamination. Guidelines for septic tank placement and maintenance ensure effective waste disposal and minimize health risks. Effective waste collection systems, including UGD and septic tanks, mitigate environmental pollution and protect water sources. Community education on proper waste disposal practices fosters sustainable sanitation solutions.

The table 4.2 provides detailed insights into the characteristics of toilets across three districts: Ariyalur, Perambalur, and Nilgiris. It includes data on the duration of toilet construction, sources of finance for construction, septic tank attachment, distance from water sources, and waste collection methods. The majority of toilets were constructed within 1-10 years across all districts, with varying percentages reflecting ongoing infrastructure development efforts. Notably, a significant number of respondents did not specify the duration of construction (NA category) that means they don't have Toilet facilities in their houses.

The people's Own funding is the predominant source across all districts, indicating individual household investment in sanitation infrastructure. Some respondents received financial support from government schemes, NGOs, or other sources, though these contributions are relatively minor in comparison. Majority of toilets are attached to septic tanks in all districts, which is crucial for proper waste management and sanitation. A notable percentage of respondents did not specify septic tank attachment status (NA category). Variability exists in the distance between septic tanks and water sources across districts, with efforts needed to ensure adequate separation to prevent contamination. Underground Drainage (UGD) systems and septic tanks are the primary methods of waste collection across all districts. Other methods like direct discharge to drainage, open places, or water bodies are less common but still present in smaller percentages.

The data reveals significant progress in sanitation infrastructure across Ariyalur, Perambalur, and Nilgiris, with a high percentage of households having access to toilets. Most toilets are constructed within the past decade, highlighting ongoing efforts in sanitation development. The predominant use of septic tanks and varied waste collection methods underscores the importance of proper waste management practices for environmental and public health.

4.2 Characteristics of Individual Toilets

Table 4.2

District Wise Percentage distribution of Characteristics of the Toilet

Characteristics of Toilets	**Ariyalur**		**Perambalur**		**Nilgiris**	
	N	%	N	%	N	%
Duration of Toilet construction						
Below 1 year	-	-	253	20.6	425	19.1
1-5 years	350	31.9	257	20.9	434	19.5

6-10 years	313	28.5	277	22.5	466	20.9
11-15 years	136	12.4	259	21.2	557	25.0
<15years	49	4.5	-	-	-	-
NA	249	22.7	183	14.9	344	15.5
Total	1097	100.0	1229	100.0	2226	100.
Source of finance for Toilet construction						
No Idea	-	-	253	20.6	425	19.1
Own	828	75.5	733	59.6	1339	60.2
Slum Board	17	1.5	-	-	-	-
Govt & NGO	3	0.3	60	4.9	118	5.3
NA	249	22.7	183	14.9	344	15.5
Total	1097	100.0	1229	100.0	2226	100.
Septic attached						
Yes	767	69.9	371	30.2	1066	47.9
No	81	7.4	675	54.9	816	36.7
NA	249	22.7	183	14.9	344	15.5
Total	1097	100.0	1229	100.0	2226	100.
Distance between septic Tank and Water Source						
Below 5m	-	-	675	54.9	816	36.7
5 m	366	33.4	112	9.1	242	10.9
10 m	189	17.2	86	7.0	186	8.4
15 m	136	12.4	42	3.4	283	12.7
Above 16 m	157	14.3	131	10.7	355	15.9
NA	249	22.7	183	14.9	344	15.5
Total	1097	100.0	1229	100.0	2226	100.
Waste collection method						
UGD	127	11.6	433	35.2	530	23.8
Septic tank	713	65.0	395	32.1	1247	56.0
Discharge to drainage	2	0.2	201	16.4	84	3.8

Discharge in open place	4	0.4	9	0.7	3	0.1
Discharge in water bodies	1	0.1	1	0.1	9	0.4
Bio-septic tank	-	-	1	0.1	9	0.4
Leach pit	1	0.1	6	0.5	-	-
NA	249	22.7	183	14.9	344	15.5

Sanitation facilities play a critical role in public health and environmental sustainability. Issues such as septic tank overflow, broken infrastructure, and pipe blockages can lead to health hazards and environmental pollution. Effective maintenance and timely repairs are essential to prevent these problems. Regular inspection and maintenance of toilets and associated infrastructure are key to preventing issues like septic tank overflow and pipe blockages. Training programs for homeowners on basic maintenance can improve sustainability. Ensuring robust construction standards and quality materials for toilets can mitigate issues like broken basins and stinking facilities. Proper design and installation reduce maintenance needs and prolong infrastructure lifespan.

Table 4.3 presents data on various issues related to individual toilets across three districts: Ariyalur, Perambalur, and Nilgiris. The responses are categorized into different issues faced by respondents with their toilets: Frequent Overflow of Septic Tank, Broken and Stinking Basin, Blockage of Pipes, No Issues. Overall, the majority of respondents in Ariyalur, Perambalur, and Nilgiris reported no significant issues with their individual toilets. However, a small percentage faced challenges such as septic tank overflow, broken basins, and pipe blockages. Addressing these issues is crucial to ensure continued functionality and hygiene of sanitation facilities.

Table 4.3

District wise Multi Responses of the Individual Toilet Issues

	Ariyalur	Perambalur	Nilgiris

Multi Responses Issues[a]	N	%	N	%	N	%
Frequent overflow the septic tank	57	6.7	19	3.2	104	7.8
Broken and stinks the basin	6	0.7	18	3.0	64	4.8
Blockage pipe	5	0.6	21	3.5	74	5.5
No Issues	1323	98.7	599	99.5	800	94.3
a-Dichotomy group tabulated at value 1 Percentage and Totals are based on respondents						

The most serious documented problems involve contamination of surface waters and ground water with disease-causing pathogens and nitrates. Other problems include excessive nitrogen discharges to sensitive coastal waters and phosphorus pollution of inland surface waters, which increases algal growth and lowers dissolved oxygen levels. Contamination of important shellfish beds and swimming beaches by pathogens is a concern in some coastal regions healthy. (EPA, 2023).

Table 4.4

District Wise Details of Individual Household Toilet Maintain

Details of Toilets	Ariyalur		Perambalur		Nilgiris	
	N	%	N	%	N	%
Toilet Cleaning						
NA	249	22.7	627	51.0	885	39.8
Once a day	712	64.9	486	39.5	979	44.0
Two days once	86	7.8	78	6.3	234	10.5
Weekly once	48	4.4	35	2.8	115	5.2
Monthly once	2	0.2	3	0.2	13	0.6

Total	1097	100.0	1229	100.0	2226	100
Toilet Maintain Cost						
Don't know	249	22.7	627	51.0	887	39.8
<=100	109	9.9	92	7.5	278	12.5
101-200	308	28.1	185	15.1	420	18.9
201-300	229	20.9	196	15.9	256	11.5
>300	202	18.4	129	10.5	385	17.3
Total	1097	100.0	1229	100.0	2226	100.0
Hand Wash Materials						
NA	249	22.7	185	15.1	355	15.9
Soap	585	53.3	593	48.3	1252	56.2
Detergent	42	3.8	4	0.3	168	7.5
Hand – liquid	200	18.2	168	13.7	378	17.0
Only water	21	1.9	279	22.7	73	3.3
Total	1097	100.0	1229	100.0	2226	100.0
Tank Cleaning						
NA	249	22.7	624	50.8	885	39.8
Annually	302	27.5	322	26.2	279	12.5
Once in 2 years	36	3.3	47	3.8	26	1.2
Beyond 2 years	49	4.5	71	5.8	163	7.3
Never	456	41.6	144	11.7	658	29.6
Don't know	5	0.5	21	1.7	215	9.7

Total	1097	100.0	1229	100.0	2226	100.0
Who cleans the septic tank						
NA	249	22.7	180	14.6	344	15.5
Self	197	18.0	457	37.2	674	30.3
Govt	30	2.7	443	36.0	440	19.8
NGO	-	-	22	1.8	4	0.2
Private agencies	155	14.1	127	10.3%	764	34.3
Never	466	42.5	-	-	-	-
Total	1097	100.0	1229	100.0	2226	100.0
The cost is to pick up the septic tank waste in Rs.						
Dont' Know	357	32.5	297	24.2	753	33.8
>1000	118	10.8	247	20.1	237	10.6
1001-2000	62	5.7	273	22.2	278	12.5
2001-5000	268	24.4	141	11.5	219	9.8
5001-10000	33	3.0	65	5.3	195	8.8
<10000	10	0.9	26	2.1	200	9.0
NA	249	22.7	180	14.6	344	15.5
Total	1097	100.0	1229	100.0	2226	100.0

In Ariyalur, Perambalur, and Nilgiris districts, the frequency of toilet cleaning varies significantly. In Ariyalur, the majority (64.9%) clean toilets once a day, while in Perambalur and Nilgiris, this percentage is lower at 39.5% and 44.0%, respectively. A notable proportion in Perambalur (22.7%) and Nilgiris (10.5%) clean toilets less frequently, such as once every two days or weekly. Regarding the cost of toilet maintenance, there is a range across districts. In Ariyalur, a significant portion (28.1%) spends between 101 to 200 Rs, whereas in Perambalur and Nilgiris, smaller percentages spend in this range (15.1% and 18.9% respectively). A considerable number in Perambalur (51.0%) and Nilgiris (39.8%) are unsure of the maintenance costs. The choice of hand wash materials varies notably. Across all districts, soap is the most commonly used material (ranging from 48.3% to 56.2%), followed by hand liquid (13.7% to 18.2%). Notably, in Perambalur, a significant proportion (22.7%) uses only water, which is higher compared to other districts. Tank cleaning frequencies also vary significantly. In Ariyalur and Perambalur, a substantial number (41.6% and 27.5% respectively) clean tanks annually, whereas in Nilgiris, only 12.5% do so annually.

A significant portion in all districts (27.6% to 41.6%) never clean their tanks. The responsibility for cleaning septic tanks differs widely. In Ariyalur and Nilgiris, a notable percentage (42.5% and 36.0% respectively) never clean their septic tanks, while in Perambalur, self-cleaning is more prevalent (37.2%). Private agencies play a significant role in Nilgiris (34.3%) compared to other districts. The cost for picking up septic tank waste varies widely. A substantial number across districts (24.2% to 33.8%) are uncertain about the costs. In terms of actual costs, significant percentages in Ariyalur (24.4%) and Perambalur (22.2%) fall in the 2001-5000 Rs range, while in Nilgiris, higher percentages (33.8%) fall into the 1001-2000 Rs range. Based on these findings, it is recommended to standardize and promote regular toilet cleaning practices across all districts. Awareness programs should emphasize the importance of regular maintenance to reduce long-term costs and environmental impact. Local governments should collaborate with NGOs and private agencies to improve septic tank cleaning services, ensuring affordability and accessibility. Additionally, education campaigns should promote proper handwashing techniques using effective materials like soap to improve hygiene standards across communities.

4.3 Description of without toilets' household:

Access to sanitation facilities significantly impacts public health, dignity, and environmental sustainability. Studies emphasize the importance of community engagement, economic incentives, and cultural factors in promoting toilet adoption and usage. Involving local communities in decision-making and awareness campaigns can foster a sense of ownership and encourage participation in sanitation improvement initiatives. Providing subsidies or financial assistance for toilet construction can alleviate financial barriers and incentivize households to invest in sanitation infrastructure.

Table 4.5

District Wise Percentage Distribution of People' willingness to Build Toilet

People's Willingness	District					
	Nilgiris		Perambalur		Ariyalur	
	N	%	N	%	N	%
Yes	154	6.9	91	7.4	187	17.0
No	190	8.5	89	7.2	62	5.7
NA	1882	84.5	1049	85.4	848	77.3
Total	2226	100.0	1229	100.0	1097	100.0

Table 4.4 illustrates the willingness of households without toilets in Ariyalur, Perambalur, and Nilgiris districts to build toilets: In Nilgiris 6.9% of households expressed willingness to build toilets, in Perambalur 7.4% of households are willing to build toilets and in Ariyalur 17.0% of households showed willingness. Across all districts, 9.5% of households are willing to build toilets. The majority of households (83.0% across all districts) did not respond (NA) regarding their willingness, which could indicate uncertainty or lack of immediate plans regarding toilet construction.

As well as the table 4.5 reveals the multi response of the options of the toilet to the without toilet household. 12.6 per cent people chose the free toilet, 13.4 per cent chosen free sanitary complexes, paid toilet, 17.9 per cent chose common toilet of the line house, and 31.5 per cent chose the Open Defecation. According to this data the high proportion of the people chose the open defecation.

Table 4.6

District Wise Multi Responses of Toilet options of without Toilet household'

Multi Responses of Option of Toilet[a]	Ariyalur		Perambalur		Nilgiris	
	N	%	N	%	N	%
Free Toilet	37	15.0	55	30.2	109	32.2
Free Sanitary Complexes	-	-	40	22.0	116	34.3
Paid Toilet	-	-	44	24.2	107	31.7
Paid sanitary complex	-	-	30	20.3	155	45.9
Open Defecation	213	86.6	154	84.6	273	80.8
a-Dichotomy group tabulated at value 1 Percentage and Totals are based on respondents						

Access to adequate sanitation is crucial for public health and dignity. Studies emphasize that lack of access to toilets contributes to the spread of diseases, particularly in densely populated areas and among vulnerable populations.

Figure 4.1

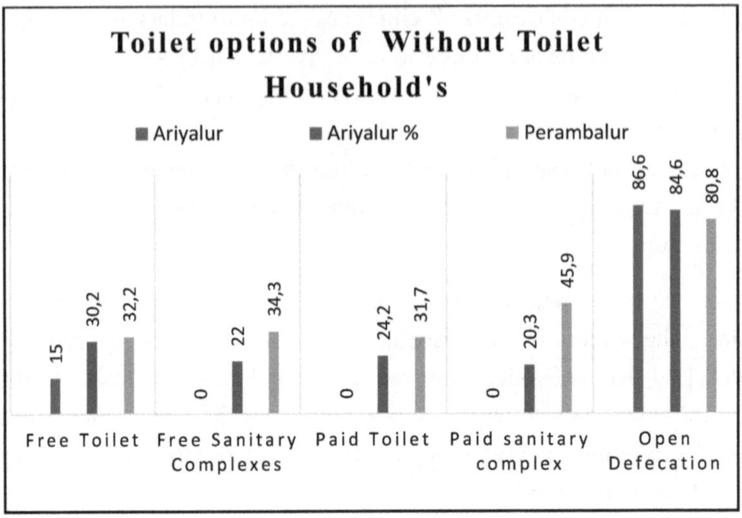

Promoting behavior change through education and community engagement can encourage adoption of improved sanitation practices. Building and maintaining sanitation facilities, including free or subsidized options, are essential to reducing open defecation rates. Table 4.5 presents the options chosen by households without toilets in Ariyalur, Perambalur, and Nilgiris districts: in Ariyalur 15.0% opt for free toilets, 86.6% resort to open defecation. In Perambalur 30.2% prefer free toilets, 84.6% choose open defecation and in Nilgiris 32.2% select free toilets, 80.8% opt for open defecation.

4.4 Description of Public Toilet:

Access to clean and safe public toilets is crucial for promoting hygiene, preventing diseases, and ensuring dignity, particularly for vulnerable groups like women and children. Studies emphasize the need for gender-sensitive sanitation facilities and accessibility features for elderly and physically handicapped individuals. Providing separate and safe sanitation facilities for women and girls can enhance their comfort and encourage regular use. Designing public toilets with features such as ramps, handrails, and spacious layouts can improve accessibility for elderly and physically challenged individuals.

Public toilets are essential infrastructure to guarantee the right to sanitation in public spaces and, in more general terms, the right to inclusive and sustainable cities. Moreover, since the equipment has a direct user interface, it is important to understand their demands and needs (Fernanda Deister Moreira, 2021). The users of public toilet were women, transgender, homeless, elderly and people with disabilities are vulnerable users (Fernanda Deister Moreira, et.al., 2021)

Table 4.7

Multi Responses of Types of persons using public toilet

Types of People[a]	Responses		Percent of Cases
	N	%	
Male	173	32.7	64.1

Female	267	50.5	98.9
Boys under 18	51	9.6	18.9
Girls under 18	29	5.5	10.7
Elderly/PH	9	1.7	3.3
Total	529	100.0%	195.9%
a-Dichotomy group tabulated at value 1			

Table 4.6 provides insights into the types of individuals using public toilets: 32.7 per cent male, 50.5 per cent female, adolescent boys 9.6 per cent 5.5 per cent girl adolescent and 1.7 per cent of elders and Physically challenges using the Public Toilets. The percentages total to 195.9%, likely due to respondents selecting multiple categories or overlapping responses.

4.5 Infrastructure defects in Public Toilet:

A Similar finding was reported in the study on "what motivates open defecation? A qualitative study from a rural setting in Nepal" (Navin Bhatt et.al., (2019) was find out the public latrines are considered more inconvenient due to long queues and considered dirty. Women were concerned regarding safety, privacy and unhygienic sanitation in the public facilities. Most of the studies mentioned the people wants to improve the toilet infrastructure. Assessment of Community and Public Toilets in Chennai study's feedback from users and communities largely focused on needing better maintenance and improvements to the physical conditions of the toilet structures (IIHS, 2022).

Table 4.8

Multi Responses of Infrastructure defects of Public Toilet

Types of defects[a]	Responses		Percent of Cases
	N	%	
Broken door/no door	117	8.0	48.5
No ramp/broken ramp	134	9.1	55.6
Limited number of seats	113	7.7	46.9

Close proximity of male-female	80	5.4	33.2
Impossible design for users	97	6.6	40.2
No hand dryer/no hand tissues	113	7.7	46.9
No Bag hanging	83	5.6	34.4
No enough space inside	81	5.5	33.6
No pipes near the toilet	152	10.3	63.1
Not suitable for children	79	5.4	32.8
Broken washbasin	77	5.2	32.0
No bathing cubicles	62	4.2	25.7
No water tub and stone paved	80	5.4	33.2
Not napkin vending machine	68	4.6	28.2
Diapers vending machine	74	5.0	30.7
No incinerators	60	4.1	24.9
Total	1470	100.0	610.0
a-Dichotomy group tabulated at value 1			

Table 4.7 multi responses of outlines various infrastructure defects in public toilets and their respective responses and percentages. Key issues identified include: Broken door/no door 8.0% of responses, affecting 48.5% of cases, No ramp/broken ramp 9.1% of responses, affecting 55.6% of cases, Limited number of seats: 7.7% of responses, affecting 46.9% of cases, Close proximity of male-female toilets 5.4% of responses, affecting 33.2% of cases, Impossible design for users 6.6% of responses, affecting 40.2% of cases, No hand dryer/no hand tissues 7.7% of responses, affecting 46.9% of cases, No pipes near the toilet 10.3% of responses, affecting 63.1% of cases and other issues Include inadequate space, lack of bathing cubicles, no napkin vending machines, and broken washbasins. The total number of responses (1470) indicates multiple issues being reported by the respondents. Access to well-maintained public sanitation facilities is crucial for public health and well-being. Defective infrastructure in public toilets can deter usage and contribute to poor hygiene practices. Studies emphasize the importance of user-friendly designs, regular maintenance, and the availability of essential amenities in public toilets. The key findings Toilets should be designed to cater to all

users, including individuals with disabilities, ensuring regular maintenance can prevent and address common defects such as broken doors and washbasins and providing necessary amenities like hand dryers, pipes, and adequate space can significantly improve user experience and hygiene. A similar study on public toilet (Yasemin Afacan, 2015) explored the physical problems with toilets are related mostly to accessibility, usability, ease of manipulation, and wayfinding. Among these, accessibility is the most critical issue.

Another similar study finding was reported in the study on "what motivates open defecation? A qualitative study from a rural setting in Nepal" (Navin Bhatt et.al., (2019) was find out the public latrines are considered more inconvenient due to long queues and considered dirty. Women were concerned regarding safety, privacy and unhygienic sanitation in the public facilities. Most of the studies mentioned the people wants to improve the toilet infrastructure. Assessment of Community and Public Toilets in Chennai study's feedback from users and communities largely focused on needing better maintenance and improvements to the physical conditions of the toilet structures (IIHS, 2022).

Table 4.9

Multi Responses of Maintenance defects of Public Toilet

Types of defects[a]	Responses		Percent of Cases
	N	%	
Insufficient water	138	8.7	60.3
No water facility	103	6.5	45.0
Unclean stored water	112	7.1	48.9
Water taps broken	111	7.0	48.5
Poor quality of bucket/mug	115	7.3	50.2
No hand washing materials	109	6.9	47.6
Unclean Toilets	116	7.4	50.7
Stinking Toilets	107	6.8	46.7
Used liquor bottle/cigarette/	68	4.3	29.7

Overflow of septic tank	127	8.0	55.5
UGD sewer breakage	53	3.4	23.1
Blockage on pipes	30	1.9	13.1
Damage due to disasters	81	5.1	35.4
Long Qudes	90	5.7	39.3
Insufficient Light	84	5.3	36.7
Long Distance	84	5.3	36.7
Fear of Gender based issues	23	1.5	10.0
Insect/animal nuisance	57	3.6	24.9
Total	1578	100.0	689.1
a - Dichotomy group tabulated at value 1			

Access to clean and functional sanitation facilities is crucial for public health and environmental sustainability. Studies emphasize the importance of maintaining adequate water supply, sanitation infrastructure, and hygiene facilities in public toilets to ensure their effective use and minimize health risks. Table 4.8 presents various infrastructure defects in public toilets and their corresponding responses and percentages across different districts. Key issues identified include: Insufficient water is reported in 8.7% of cases, affecting 60.3% of respondents. No water facility noted in 6.5% of cases, affecting 45.0% of respondents, unclean stored water Identified in 7.1% of cases, affecting 48.9% of respondents, water taps broken found in 7.0% of cases, affecting 48.5% of respondents, poor quality of bucket/mug reported in 7.3% of cases, affecting 50.2% of respondents, no hand washing materials is noted in 6.9% of cases, affecting 47.6% of respondents, unclean toilets are identified in 7.4% of cases, affecting 50.7% of respondents, Stinking toilets is found in 6.8% of cases, affecting 46.7% of respondents, Other issues are Include problems like littering with liquor bottles or cigarettes, septic tank overflows, sewer breakages, and blockages. The total number of responses (1578) highlights multiple issues being reported by users across the surveyed districts.

4.6 Open Defecation:

Open defecation contributes to public health risks, environmental pollution, and the spread of diseases such as diarrhea and cholera. Studies underscore the importance of achieving universal access to sanitation facilities to improve public health outcomes and promote dignity and safety, particularly among vulnerable populations. Studies emphasize that open defecation poses significant health risks, contributes to environmental contamination, and perpetuates social inequities. Effective interventions include community-led sanitation programs, infrastructure development, and behavior change campaigns aimed at promoting toilet use and reducing open defecation. Literature underscores that the challenges associated with open defecation contribute to public health risks, social inequalities, and environmental degradation. Effective strategies focus on improving access to sanitation facilities, enhancing safety measures, and fostering community engagement.

The available quantitative data on open defecation provides limited information regarding the motivations for open defecation (UNICEF,2018). District Wise Percentage Distribution of Open Defecation (Table 4.9) is presenting the percentage distribution of open defecation across Ariyalur, Perambalur, and Nilgiris districts. Key findings include in Ariyalur 19.4% of respondents reported practicing open defecation, Perambalu 12.5% of respondents reported practicing open defecation, Nilgiris 12.3% of respondents reported practicing open defecation. Overall, a significant portion of the population in Ariyalur (19.4%), Perambalur (12.5%), and Nilgiris (12.3%) districts still engage in open defecation despite efforts to promote sanitation and toilet usage. Cultural norms, lack of awareness, and accessibility issues contribute to the persistence of open defecation. Effective sanitation interventions include community engagement, infrastructure development, and behavior change communication. Open defecation exacerbates health risks and hinders progress toward achieving Sustainable Development Goals related to water and sanitation.

Table 4.10

District Wise Percentage distribution of Open Defecation

Open Defecation[a]	Ariyalur		Perambalur		Nilgiris	
	N	%	N	%	N	%
Yes	213	19.4	154	12.5	273	12.3
No	884	80.6	1075	87.5	1953	87.7
Total	1097	100.0	1229	100.0	2226	100.0
a-Dichotomy group tabulated at value 1 Percentage and Totals are based on respondents						

The Table 4.10 is Multi Responses of Place of Open Defecation. This table illustrates the various locations where open defecation occurs across surveyed areas, categorized by district. Key findings include: Road side/railway tracks: 13.6%, Bushes and thorns: 21.4%, Space reserved for toilet: 6.3%, Just behind the house: 9.6%, Around the water ways: 25.3%, Empty plots/fields: 23.8% . These percentages reflect the distribution of open defecation practices in different environmental settings within Ariyalur, Perambalur and Nilgiris districts.

The Table Multi Responses of Difficulties When Open Defecation (Table 4.11) This table presents the challenges encountered during open defecation in surveyed areas, categorized by specific difficulties and their percentages: Fear of snakebite/animal: Reported by 21.4% of respondents, Lack of privacy: Cited by 11.7% of respondents, Gender-based violence: Mentioned by 9.5% of respondents, Waiting for the dark: A concern for 13.0% of respondents, Distance: A challenge for 11.7% of respondents, Inconvenience due to shyness: Noted by 9.8% of respondents, Waiting for others to leave: A factor for 9.2% of respondents, Requirement for company: Indicated by 6.7% of respondents, Inconvenience for elderly and people with disabilities (PH): Reported by 7.0% of respondents.

Table 4.11

Multi Responses of Place of Open Defecation

Place of Open Defecation[a]	Responses		Percent of Cases
	N	%	

Road side/railway tracks	117	13.6	23.2
Bushes and thorns	184	21.4	36.4
Space reserved for toilet	54	6.3	10.7
Just behind the house	83	9.6	16.4
Around the water ways	218	25.3	43.2
Empty plots/fields	205	23.8	40.6
Total	861	100.0	170.5
a - Dichotomy group tabulated at value 1			

Table 4.12

Multi Responses of difficulties when Open Defecation

Place of Open Defecation[a]	Responses		Percent of Cases
	N	%	
Fear of snakebite/animal	390	21.4	77.2
Lack of privacy	214	11.7	42.4
Gender-based violence	174	9.5	34.5
Wait for the dark	237	13.0	46.9
Distance	213	11.7	42.2
Inconveniency shy	178	9.8	35.2
Waiting for the guy to leave	167	9.2	33.1
Require company	123	6.7	24.4
Inconvenient to aged & PH	128	7.0	25.3
a - Dichotomy group tabulated at value 1			

Thus, further research on open defecators can identify the physical challenges they face. If the public toilets are properly maintained and the defects in them are repaired by the government, people who do not have toilets can benefit from using them. Also, in this study, the households using the Indian style toilet and the households using the Weston style toilet can be selected to investigate their health-related problems. Because we can find if anyone has constipation and urinary problems. (Navin Bhatt et.al., 2019) was concluded in his study, the "Open defecation is either a voluntary choice or a compulsion. This choice is closely linked with personal preferences, cultural and traditional norms with special concerns for privacy for women and girls in different communities" Open defecation is common despite the presence of household toilets in the Dharmapuri district of Tamil Nadu. Toilet use is influenced by several structural and sociocultural determinants. The mission needs to shift its emphasis from toilet construction to sustainable functioning and use of toilets. (Nallathambi Yoganath & Tarun Bhatnagar, 2018). According to this data open defecation is not good option for the people. People experience various problems with open defecation. These problems affect health.

5. Environmental & Personal Hygiene

The pollution load on environmental components is creasing exponentially. The take, an artificial one formed by John Sullivan around 18th century is a boon to Ooty. Being the located at a low level, the entire rain water as well the sewage is flowing towards the lake and polluted year by year. In order to restore the quality of lake, several effective measures are taken by the Government including bioremediation by public works department. A study on the pollution effect to this lake been conducted as a project and the factors responsible for the pollution has been studied and present in the report. Necessary preventive measures and suggestion for the improvement of the solid and sewage management has been made (Subramani T, 2012). Literature emphasizes that effective solid waste management is crucial for environmental sustainability and public health. Awareness campaigns, infrastructure development, and community participation are key strategies for improving waste management practices.

5.1 Awareness of Environmental Hygiene

This chapter deals about the waste management like solid and liquid waste moreover it is including the awareness of solid waste, method of solid waste, and liquid waste also. Indian cities - especially those dealing with high population density, poverty, and inadequate sanitation – have long been susceptible to diseases. The significance of urban sanitation in India is diverse - it profoundly impacts public health, environmental sustainability, and quality of life. As India's urban population grows and urbanization progresses at full tilt, ensuring clean and hygienic living conditions becomes challenging. Inadequate urban sanitation can propagate waterborne diseases, inflate healthcare costs, and deteriorate the quality of life for millions. It also poses severe environmental threats like water pollution and ecosystem degradation. In countries with massive populations, the existing infrastructure is often overburdened, leading to frequent breakdowns and escalating maintenance costs (Ankitha Gattupalli, 2023).

Table 5.1

District Wise percentage distribution of Awareness of solid waste Management

Details of Waste Management[a]	**Ariyalur**		**Perambalur**		**Nilgiris**	
	N	%	N	%	N	%
Knowledge about solid waste	895	81.6	727	59.2	2003	90.0
Have Trash at home	936	85.3	1067	86.8	2030	91.2
Near Public Trash	299	27.3	111	9.0	1059	47.6
Separate the waste	844	76.9	715	58.2	1912	85.9
a-Dichotomy group tabulated at value 1, % and Totals are based on respondents						

Literature emphasizes that effective solid waste management is crucial for environmental sustainability and public health. Awareness campaigns, infrastructure development, and community participation are key strategies for improving waste management practices. The Table 5.1 presents the awareness levels and practices related to solid waste management across three districts: **Knowledge about solid waste**: Ariyalur has the highest awareness at 81.6%, followed by Nilgiris at 90.0%, while Perambalur stands at 59.2%. **Have Trash at home**: High percentages across all districts, with Nilgiris leading at 91.2%. **Near Public Trash**: Ariyalur and Nilgiris show moderate awareness (27.3% and 47.6%, respectively), whereas Perambalur has lower awareness (9.0%). **Separate the waste**: Ariyalur and Nilgiris exhibit higher rates of waste separation (76.9% and 85.9%, respectively) compared to Perambalur (58.2%). Awareness about solid waste is relatively high, indicating potential for effective interventions. Majority of households have trash at home, highlighting the widespread need for waste collection services.

Disparities exist in waste segregation practices, suggesting a need for targeted education and infrastructure improvements.

5.2 Solid Waste Disposal

Research underscores the importance of effective waste disposal methods in promoting public health and environmental sustainability. Efficient waste collection and disposal systems reduce pollution, mitigate health risks, and conserve resources. The Nilgiris municipal administration deployed the works to collect door to door collection of degradable and bio degradable waste in each household which will be further making an organic manure. The wastewater generated from the hotels and cottages can be treated which will be reused for gardening purposes which has creates the opportunity cost of using ground water. (Nilgiris District Diagnostic Studies-DDS, 2018).

Table 5.2
District Wise Multi Response of Solid Waste Disposal Method

Solid waste Disposal Method[a]	Ariyalur		Perambalur		Nilgiris	
	N	%	N	%	N	%
Disposal in public dustbins	44	4.0	114	9.3	372	16.7
Dump in backyards/compound	138	12.6	69	5.6	120	5.4
Used for compost preparation	53	4.8	28	2.3	46	2.1
Throw in open space, street end	67	6.1	76	6.2	93	4.2
Collected by sanitation workers	1010	92.1	1140	92.8	1934	86.9
Garbage collection & incineration	58	5.3	32	2.6	66	3.0
Burry	24	2.2	11	0.9	92	4.1
Throwing in open sewer	15	1.4	12	1.0	15	0.7
Dump in the road side and burn	39	3.6	22	1.8	11	0.5

a-Dichotomy group tabulated at value 1 Percentage and Totals are based on respondents

The Table 5.2 illustrates the various methods of solid waste disposal across three districts. Higher in Nilgiris (16.7%) compared to Ariyalur (4.0%) and Perambalur (9.3%) **Disposal the solid waste in public dustbins**. More prevalent in Ariyalur (12.6%) than in Perambalur (5.6%) and Nilgiris (5.4%) **Dumping the waste in backyards/compound.** Lowest overall, with Ariyalur at 4.8%, Perambalur at 2.3%, and Nilgiris at 2.1% use the waste for compost preparation. Similar across districts, with Ariyalur at 6.1%, Perambalur at 6.2%, and Nilgiris at 4.2% throw the wastage in open space, street end etc., Highest response in all districts, particularly prominent in Ariyalur (92.1%), Perambalur (92.8%), and Nilgiris (86.9%) collected the wastage by the sanitation workers. More prevalent in Ariyalur (5.3%) compared to Perambalur (2.6%) and Nilgiris (3.0%) garbage collection and incineration. More common in Nilgiris (4.1%) than in Ariyalur (2.2%) and Perambalur (0.9%) buried the wastage. Minimal across districts, with Ariyalur at 1.4%, Perambalur at 1.0%, and Nilgiris at 0.7% throwing the wastage in open sewer. And more prevalent in Ariyalur (3.6%) compared to Perambalur (1.8%) and Nilgiris (0.5%) dump the wastage in the roadside and burn. High reliance on sanitation workers for waste collection underscores existing infrastructure but also points to challenges in informal settlements. Instances of improper disposal methods like burning and open dumping suggest gaps in public awareness and enforcement of waste management regulations. Variations in waste disposal methods between districts necessitate context-specific interventions tailored to local needs.

5.3 Environmental Hygiene and Sanitation Workers:

Effective waste collection is crucial for maintaining public health and environmental cleanliness. Regular waste removal reduces the risk of disease transmission, minimizes environmental pollution, and enhances overall community well-being. The Table 5.3 presents the frequency of waste collection by sanitation workers across three districts: Most common in Nilgiris (88.1%), Perambalur (88.9%), and less so in Ariyalur (71.9%) collect the wastage everyday by the sanitation workers. Minimal in Nilgiris (3.3%) and Perambalur (3.9%), higher in Ariyalur (10.2%) collect once in a week, more frequent in Ariyalur (13.2%), moderate in Nilgiris (6.4%) and Perambalur (3.5%) collect twice a week, and Lowest overall, with Nilgiris at 2.2%, Perambalur at 3.7%, and Ariyalur at 4.7% no service of sanitation workers.

Table 5.3

District Wise percentage distribution of Sanitation Workers' Method of Collecting waste

Frequency of waste collection by Sanitation workers	District					
	Nilgiris		Perambalur		Ariyalur	
	N	%	N	%	N	%
Everyday	967	88.1	1093	88.9	1601	71.9
Once in a week	36	3.3	48	3.9	226	10.2
Twice in a week	70	6.4	43	3.5	294	13.2
No Service	24	2.2	45	3.7	105	4.7
Total	2226	100.0	1229	100.0	1097	100

High rates of daily waste collection in most areas reflect robust sanitation services but also highlight areas where service gaps exist. Areas with less frequent collection may experience challenges in waste accumulation and sanitation-related issues and instances of no service underscore infrastructure challenges and the need for targeted interventions to improve waste management.

5.4 Liquid Waste Disposal

Efficient liquid waste management is crucial for environmental sustainability and public health. Proper disposal methods reduce contamination risks to water sources, soil, and air, thereby preventing disease outbreaks and environmental degradation. Table 5.4 illustrates the various methods used for liquid waste disposal across three districts: Utilized Soak Pit by 22.7% in Ariyalur, 9.9% in Perambalur, and 11.1% in Nilgiris, Predominantly connected the liquid wastage to drainage in Perambalur (81.9%), followed by Nilgiris (68.7%) and Ariyalur (40.4%), minimal treat and reuse the liquid waste overall, with 0.5% in Ariyalur, 0.2% in Perambalur, and 2.0% in Nilgiris, Seen in 16.5% of Ariyalur, 4.8% of Perambalur, and 10.6% of Nilgiris diverted the liquid waste to their garden, use for home gardening is common in Ariyalur (31.7%), less so in Nilgiris (12.4%) and Perambalur (6.2%), people practices the waste water pouring into the open place by 23.6% in Ariyalur, 6.9% in Perambalur, and 13.0% in Nilgiris, people dumping on the waste water in water bodies like river, lake, well, stream etc. 6.7% in Ariyalur, 2.0% in Perambalur, and 2.9% in Nilgiris and minimal per centage of the people using the waste water to farming in Ariyalur 1.1 %, 0.7% in Perambalur, and 4.5% in Nilgiris.

l methods indicate varied levels of awareness, infrastructure, and cultural practices. Practices such as pouring into the open and dumping in water bodies pose risks to ecosystems and public health. Promotion of sustainable practices like reuse and safe disposal can mitigate environmental impacts.

Table 5.4
District wise Multi Responses of the Liquid Waste disposal method

Liquid waste Disposal Method[a]	Ariyalur		Perambalur		Nilgiris	
	N	%	N	%	N	%
Soak pit	249	22.7	122	9.9	246	11.1
Connected to drainage	443	40.4	1007	81.9	1525	68.7
Treat and reuse	5	0.5	2	0.2	45	2.0
Diverted to garden	181	16.5	59	4.8	235	10.6
Use for home gardening	348	31.7	76	6.2	276	12.4

Pouring in to the open	259	23.6	85	6.9	289	13.0
Dumping on water bodies	73	6.7	24	2.0	65	2.9
Connect to the farm	12	1.1	9	0.7	101	4.5
a-Dichotomy group tabulated at value 1 Percentage and Totals are based on respondents						

Figure 5.1

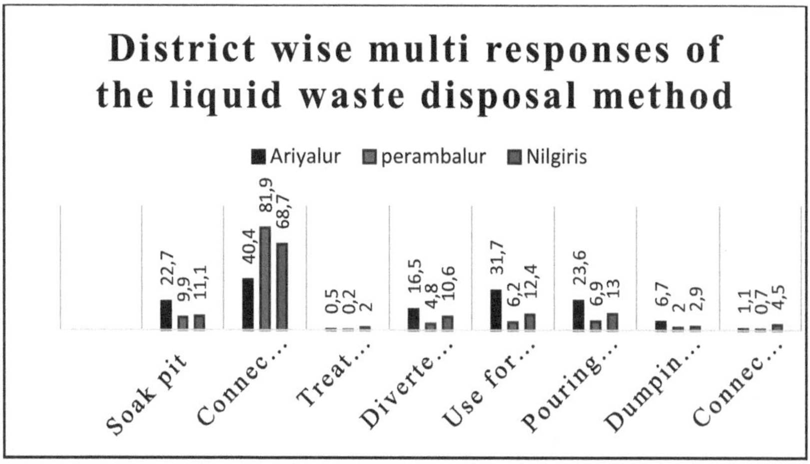

The population density in the city is very high and groundwater is polluted due to their unsafe discharge of waste water in public places, agricultural land, river, pond, lake etc. It indicates that the livelihood of a city dweller will suffer due to lack of water.

5.5 Kitchen wastage disposal

Efficient management of kitchen waste is crucial for reducing environmental pollution and promoting sustainable practices. Proper disposal methods not only mitigate health risks but also contribute to resource conservation and community hygiene. Unsafe discharge of domestic waste water by people in urban areas can create will various health problems. For example, in Perambalur, Arumbavoor, a town panchayat, during a discussion survey, it was mentioned that the wells and rivers there are now spoiled. The main reason for this is the mixing of waste water in the rivers there. Domestic sewage also pollutes open water, just as sewage in industrialized cities mixes with rivers and ponds. There is an urgent need to rectify this. And it is necessary to carry out a study on this. Especially, it is necessary to find out the quality of underground water, drinking water, all there with the help of laboratory. A Vietnam study (Hoang Minh Giang, ,2017) on Development Methodology of Sustainable Solid Waste Management System is concluded the municipal waste composition of the city had food waste as the largest proportion (42%) and hazardous waste as the smallest contributor (less than 1%). Total biodegradable waste (food and yard trimmings) was approximately 53%. Combustible waste was the second significant 91 component of about 16% while other recyclable contributed about 20% of the municipal waste composition. Similar study on wage management (Ruth Maxine Flores et.al., 2017) was mentioned the improper waste disposal is the major problem of the residents in Mojon Tampoy. And these problems were defined as incinerating, throwing their garbage on the well and the lack of garbage collection. These problems could harm not only the residents but also the environment itself.

Table 5.5 presents the various methods used for disposing of kitchen waste across three districts: Commonly kitchen waste dumping in public dustbin in Perambalur (53.2%) and Nilgiris (51.3%), less prevalent in Ariyalur (20.8%). Most frequent dumping the kitchen waste in field in Ariyalur (37.9%), less so in Perambalur (21.6%) and Nilgiris (15.5%). The kitchen waste dump in backyards in 32.4% of Ariyalur, 21.7% of Perambalur, and 30.3% of Nilgiris. Some per cent utilized the waste to use to manure pit in farm 15.9% in Ariyalur, 3.2% in Perambalur, and 8.0% in Nilgiris. **Composting** the waste in Ariyalur (29.8%), less common in Perambalur (10.2%) and Nilgiris (13.6%). **Waste Dump in Surface Water Bodies** by 2.4% in Ariyalur, 3.8% in Perambalur, and 7.0% in Nilgiris. Significant disparities in disposal methods highlight differences in infrastructure, cultural norms, and awareness levels. Practices such as dumping in water bodies pose risks to aquatic ecosystems and public health. Promoting composting and manure use can enhance soil fertility and reduce organic waste volume.

Table 5.5

District Wise Multi Responses of Dispose of Kitchen Waste

Kitchen waste Disposal Method[a]	Ariyalur		Perambalur		Nilgiris	
	N	%	N	%	N	%
Dumping public dustbin	228	20.8	654	53.2	1067	51.3
Dumping in field	415	37.9	256	21.6	322	15.5
Dump in backyards	355	32.4	267	21.7	630	30.3
Use to manure pit in farm	174	15.9	39	3.2	167	8.0
Composting	326	29.8	125	10.2	283	13.6
Dump in surface water bodies	26	2.4	47	3.8	145	7.0
a-Dichotomy group tabulated at value 1 Percentage and Totals are based on respondents						

5.6 personal Hygiene

his chapter deals about the hygiene of people, like hygiene in preparing food, personal hygiene and menstrual hygiene. Hygienic practices in food handling are critical for preventing foodborne illnesses and maintaining public health. Properly covering cooked food prevents contamination from insects, dust, and airborne pathogens, thereby reducing the risk of food spoilage and disease transmission. Hand washing is recognized as a fundamental hygiene practice that prevents the spread of infectious diseases, especially in settings where sanitation infrastructure may be inadequate. Studies emphasize the importance of consistent hand washing at critical times to reduce the transmission of pathogens and improve public health outcomes.

5.7 Hygienic practice during food preparation

Table 6.1 illustrates the methods used to keep cooked food hygienically across three districts: Practiced extensively in all districts, with high percentages: Ariyalur (99.3%), Perambalur (95.3%), and Nilgiris (98.3%)

fully covered the cooked food, Reported by a small percentage in Ariyalur (6.0%) and Nilgiris (11.1%), and very low in Perambalur (1.3%) half covered the cooked food, Minimal instances in Ariyalur (6.0%), Perambalur (0.6%), and Nilgiris (2.8%) not closed the cooked food. The high percentage of fully covered cooked food reflects awareness and adherence to hygiene standards. Instances of half-covered or uncovered food suggest opportunities for education and reinforcement of hygienic practices. Improperly covered food can lead to foodborne illnesses, impacting community health and well-being

Table 6.1

District Wise Multi Responses of Keep cooked food in Hygienically

Keep cooked food in Hygienically [a]	Ariyalur		Perambalur		Nilgiris	
	N	%	N	%	N	%
Fully Covered	1257	99.3	1171	95.3	2188	98.3
Half covered	66	6.0	16	1.3	246	11.1
Not closed	66	6.0	11	0.6	62	2.8
a-Dichotomy group tabulated at value 1 Percentage and Totals are based on respondents						

Hygienic practices during food preparation are crucial for preventing foodborne diseases and ensuring food safety. Studies emphasize the importance of proper hand hygiene, cleanliness of cooking utensils, and avoiding contamination from sick individuals or unclean surfaces.

Table 6.2

District Wise Multi Responses of Hygienic practice of during Food Preparation

Keeping cooked food in Hygienically [a]	Ariyalur		Perambalur		Nilgiris	
	N	%	N	%	N	%

Keeping fingers & nails clean	725	66.1	953	77.5	1715	77.0
Wearing a hat/tying hair tightly	185	16.9	174	14.2	599	26.9
Avoid cooking while sick	478	43.6	377	30.7	1042	46.8
Wash food items before cooking	811	73.9	674	54.8	1377	61.9
Discard cooked waste food	564	51.4	385	31.3	972	43.7
Wash hands thoroughly	595	54.2	503	40.9	1084	48.7
If have wound avoid cook	348	31.7	248	20.2	591	26.5
No cook when sneezing & cough	262	23.9	242	19.7	508	22.8
Wear Armor and gloves	217	19.8	151	12.3	248	11.1
Washing dishes regularly	778	70.9	888	72.3	1711	76.9
a-Dichotomy group tabulated at value 1 Percentage and Totals are based on respondents						

Table 6.2 presents the various hygienic practices followed during food preparation across three districts: Majority of respondents keeping fingures and nails clean in Ariyalur (66.1%), Perambalur (77.5%), and Nilgiris (77.0%), Significant practice in Nilgiris (26.9%) compared to Ariyalur (16.9%) and Perambalur (14.2%) wearing a hat/tying hair tightly, followed by a considerable proportion in Nilgiris (46.8%) and Ariyalur (43.6%), and less in Perambalur (30.7%) avoid cooking while sick, High compliance across all districts: Ariyalur (73.9%), Perambalur (54.8%), and Nilgiris (61.9%) wash food items before cooking, Practiced widely in Ariyalur (51.4%), Nilgiris (43.7%), and less in Perambalur (31.3%) discard cooked waste food, commonly practiced in Ariyalur (54.2%), Nilgiris (48.7%), and less in Perambalur (40.9%) **Wash hands thoroughly,** Followed by a significant proportion in Ariyalur (31.7%),

Nilgiris (26.5%), and less in Perambalur (20.2%) if have wound avoided cooking, practiced in varying degrees across districts: Ariyalur (23.9%), Nilgiris (22.8%), and Perambalur (19.7%) no cooking when sneezing and coughing, more commonly practiced in Ariyalur (19.8%), less in Perambalur (12.3%) and Nilgiris (11.1%) wear apron and gloves during cook, High compliance across all districts: Ariyalur (70.9%), Perambalur (72.3%), and Nilgiris (76.9%) washing dishes regularly. Practices like washing food items and washing hands are generally well-practiced, indicating awareness and possibly ingrained cultural norms. Practices such as wearing protective gear and avoiding cooking while sick show room for improvement, suggesting targeted education and awareness campaigns. Adherence to these practices directly impacts community health, reducing the risk of foodborne illnesses and promoting overall well-being.

5.8 Hand wash Habit

The table 6.3 multiple responses set reveals about the hand wash habit, the multiple set have thirteen variables, among them above the ten per cent of people having hand wash habit before and after eating food, 7.7 per cent before handling water directly, 11.1 per cent after using toilet and 11.0 per cent of the people after clean kids' and elders' stools.

Table 6.3 provides insights into the hand washing habits across various scenarios in three districts, practiced universally across all districts: Ariyalur (100.0%), Perambalur (100.0%), Nilgiris (92.7%) washed hand before take food, widely practiced in all districts: Ariyalur (100.0%), Perambalur (100.0%), Nilgiris (96.1%) after ate food only, commonly followed in Ariyalur (100.0%), Perambalur (100.0%), and less in Nilgiris (68.0%) hand wash before handling water directly, nearly universal practice across all districts: Ariyalur (100.0%), Perambalur (100.0%), Nilgiris (97.3%) hand wash after using toilet, significant practice in Perambalur (85.1%), Ariyalur (77.3%), and less in Nilgiris (66.5%) hand habit after maintenance of toilet, followed more in Perambalur (37.5%) and less in Ariyalur (29.4%) and Nilgiris (15.5%) hand wash before changing napkins, commonly followed in Ariyalur (100.0%), Perambalur (37.5%), and Nilgiris (22.4%) after changing sanitary napkins, practiced universally across all districts: Ariyalur (100.0%), Perambalur (100.0%),

Nilgiris (96.8%) **after cleaning kids' and elders' stools**, widely practiced in Ariyalur (100.0%), Perambalur (100.0%), and less in Nilgiris (63.1%) **after cleaning the house**, commonly followed in Ariyalur (100.0%), Perambalur (100.0%), Nilgiris (64.0%) **Returning from field/workplace**, widely practiced in Ariyalur (100.0%), Perambalur (100.0%), and less in Nilgiris (62.4%) **Before and after cooking food,** commonly followed in Ariyalur (100.0%), Perambalur (100.0%), Nilgiris (58.4%), **before serving meals,** significant practice in Nilgiris (56.9%), Ariyalur (12.9%), and less in Perambalur (0.0%) **after lighting lamp for worship.**

Table 6.3

District Wise Multi Response of Hand wash Habit

Hand wash Habit	Ariyalur		Perambalur		Nilgiris	
	N	%	N	%	N	%
Before food	1097	100.0	1229	100.0	2063	92.7
After food only	1097	100.0	1229	100.0	2139	96.1
Before handling water directly	1097	100.0	1229	100.0	1513	68.0
After using toilet	1097	100.0	1229	100.0	2167	97.3
After maintenance of toilet	848	77.3	1046	85.1	1480	66.5
Before change napkins	323	29.4	461	37.5	346	15.5
After changing sanitary napkins	1097	100.0	461	37.5	498	22.4
After clean kids & elder' stools	1097	100.0	1229	100.0	2155	96.8
After cleaning the house	1097	100.0	1229	100.0	1405	63.1
Returning from field/workplace	1097	100.0	1229	100.0	1424	64.0
Before and after cook food	1097	100.0	1229	100.0	1389	62.4

Before serving meals	1097	100.0	1229	100.0	1229	58.4
After light lamp to worship	141	12.9	1229	100.0	1267	56.9
a-Dichotomy group tabulated at value 1 Percentage and Totals are based on respondents						

Practices like hand washing before and after meals, and after using the toilet, are widely practiced across all districts, some practices, such as hand washing before handling water directly or after specific household tasks, show variability across districts and practices like hand washing after lighting a lamp for worship highlight cultural influences on hygiene behaviors. The hand wash habit is very low during the change the sanitary napkins. Statistics show that very few percentages wash their hands after changing sanitary napkin. This is much, much lower than the percentage of hand washing at other times. This shows that they pay very little attention to their cleanliness. If we carry out a study on this, we can know how women follow their cleanliness during menstruation and what are the other factors related to it. That means place of sanitary nappy changing or privacy or toilet facility, water facility, etc. can be known.

5.9 Bathing:

Bathing frequency is influenced by cultural norms, climate, and access to water and sanitation facilities. Regular bathing is essential for personal hygiene, particularly in tropical climates where heat and humidity can increase the risk of skin infections and other health issues.

Table 6.4 presents the frequency of bathing habits across three districts: Most common once a day bathing frequency in Ariyalur (88.4%), Perambalur (81.2%), and less in Nilgiris (58.4%), followed by a significant portion twice a day bathing in Perambalur (18.0%), Ariyalur (10.1%), and least in Nilgiris (9.5%), minimal practice once a week bathin in Ariyalur (0.5%), Perambalur (0.2%), and more in Nilgiris (3.4%), practiced less twice a week bathing in Ariyalur (0.9%), Perambalur (0.6%), and significantly more in Nilgiris (28.1%) and not reported in Ariyalur and Perambalur, but reported by a small percentage in Nilgiris (0.6%) bathing beyond the week.

Table 6.4

District Wise Regularity of bathing Habit

Regularity of bathing	Ariyalur		Perambalur		Nilgiris	
	N	%	N	%	N	%
Once a day	970	88.4	998	81.2	1300	58.4
Twice a day	111	10.1	221	18.0	211	9.5
Once a week	6	0.5	3	0.2	76	3.4
Twice a week	10	0.9	7	0.6	626	28.1
Beyond the week	-	-	-	-	13	0.6

Access to bathing facilities is crucial for maintaining personal hygiene and preventing waterborne diseases. Household bathrooms and integrated bathroom-toilet facilities are standard indicators of sanitation access, reflecting improved living standards and health outcomes. Open spaces for bathing may indicate inadequate sanitation infrastructure or cultural preferences for communal bathing practices.

Table 6.5 shows the distribution of bathing spaces across three districts: predominantly used the household bathrooms in Ariyalur (49.0%), Perambalur (62.7%), and Nilgiris (68.5%), also commonly used bathroom along with toilet at home, though less prevalent, in Ariyalur (38.5%), Perambalur (28.3%), and Nilgiris (25.7%), used the open space at home by a minority in Ariyalur (10.2%), Perambalur (6.8%), and Nilgiris (4.4%), negligible use the public sanitary complexes in Perambalur (0.2%) and Nilgiris (0.2%), very rarely used banks of river, lake, well and near public tap across districts, found in small numbers in all districts used temporary bathing cubicles, with slightly higher prevalence in Ariyalur (1.4%). Predominant across all districts, indicating widespread access to private bathing facilities, common but less prevalent, suggesting variations in household infrastructure, notable in Ariyalur and Perambalur, highlighting potential gaps in household sanitation access and Minimal presence, indicating a need for expanded public infrastructure.

Bathing habits during menstruation are influenced by cultural practices, access to sanitation facilities, and personal hygiene beliefs. Regular bathing is essential for maintaining hygiene and comfort during menstrual periods. The variation in bathing frequencies reflects individual preferences and local cultural norms regarding menstrual hygiene management.

Table 6.6 presents the bathing habits of individuals during menstruation across three districts: Represents respondents who are male and thus not applicable to this category, prevalent in Ariyalur (52.4%), Perambalur (18.6%), and Nilgiris (40.0%). Common bathing frequency during menstruation in Ariyalur (31.7%), Perambalur (36.3%), and Nilgiris (28.4%). A lesser but notable frequency in Ariyalur (4.5%), Perambalur (9.0%), and Nilgiris (2.6%). Occasional practice observed in Ariyalur (1.7%), Perambalur (0.6%), and Nilgiris (0.8%). Respondents who have reached menopause, prevalent in Perambalur (27.2%) and Nilgiris (22.3%). Majority across all districts, highlighting the gender-specific nature of the survey question. Common practice, indicating regular hygiene practices during menstruation. Less common but observed, suggesting personal preferences or specific cultural practices. Occasional practice, likely influenced by personal comfort and hygiene needs. Different bathing habits, potentially reflecting changes in personal hygiene routines post-menopause

Table 6.5

District Wise Percentage Distribution of Bathing space

Bathing Space	Ariyalur		Perambalur		Nilgiris	
	N	%	N	%	N	%
Household bathrooms	538	49.0	771	62.7	1524	68.5
Bathroom along with toilet at home	422	38.5	348	28.3	572	25.7
Open space at home	112	10.2	83	6.8	98	4.4
Public sanitary complexes	-	-	3	0.2	4	0.2
Banks of river/lakes/well	7	0.6	2	0.2	2	0.1
Near public tap	3	0.3	2	0.2	-	-

Temporary bathing cubicles	15	1.4	20	1.6	26	1.2

Table 6.6

District Wise Bathing Habits During menstruation

Bathing Space	Ariyalur		Perambalur		Nilgiris	
	N	%	N	%	N	%
NA(Male)	575	52.4	228	18.6	890	40.0
Once a day	348	31.7	446	36.3	632	28.4
Twice a day	49	4.5	110	9.0	58	2.6
End of the menstrual flow	19	1.7	7	0.6	18	0.8
Menopause	-	-	334	27.2	496	22.3

Table 6.7

District Wise Multi Response of Benefits Soap Use

Benefits of Soap Use	Ariyalur		Perambalur		Nilgiris	
	N	%	N	%	N	%
Healthy practice	991	90.4	853	69.4	1670	75.0
Cleanliness practice	713	65.1	751	61.1	1528	68.6
Avoid infections	609	55.6	610	49.6	1466	65.9
To remove dirt	558	50.9	805	65.5	1241	55.8
a-Dichotomy group tabulated at value 1 Percentage and Totals are based on respondents						

The table 6.6 reveals about the during menstruation bathing habits, 28.4 per cent of women take bath once a day, 2.6 per cent take twice a day, 6.7 per cent of women take bath end of their menstrual flow. The table 6.7 shows about the awareness of the soap using, 28.3 per cent of thought the about the soap using is healthy practice and wise habits, 25.9 per cent told soap is cleanliness practice, 24.8 per cent thought avoid infection, 21.0 per cent thought remove the dirty.

Table 6.7 presents the perceived benefits of using soap across three districts. Acknowledged by a significant majority in Ariyalur (90.4%), Perambalur (69.4%), and Nilgiris (75.0%) as a beneficial practice for maintaining health. Identified by respondents in Ariyalur (65.1%), Perambalur (61.1%), and Nilgiris (68.6%) as contributing to personal cleanliness. Seen as important in preventing infections by respondents in Ariyalur (55.6%), Perambalur (49.6%), and Nilgiris (65.9%). And acknowledged in Ariyalur (50.9%), Perambalur (65.5%), and Nilgiris (55.8%) as effective for removing dirt.

Table 6.8

District Wise Percentage Distribution of Personal Hygiene

Personal Hygiene	**Ariyalur**		**Perambalur**		**Nilgiris**	
	N	%	N	%	N	%
Brushing Teeth						
Once a day	4056	96.3	1096	89.2	2090	93.9
Twice a day	41	3.1	133	10.8	136	6.1
Nail Cutting						
Once in a week	343	31.3	584	47.5	714	32.1
Monthly Once	205	18.7	67	5.5	329	14.8
Whenever necessary	549	50.0	578	47.0	1183	53.1
Hand Wash						
Every one hour	18	1.6	44	3.6	86	3.9
Two hours once	35	3.2	176	14.3	74	3.3
Whenever necessary	1044	95.2	1009	82.1	2066	92.8
Hair Wash						
Once a day	725	66.1	715	58.2	756	34.0
Twice a day	121	11.0	193	15.7	208	9.3

Weekly once	95	8.7	154	12.5	500	22.5
Weekly twice	156	14.2	167	13.6	762	34.2
Cloth Wash						
Daily	1010	92.1	1038	84.5	1219	54.8
Once a day	57	5.2	78	6.3	412	18.5
Twice a day	30	2.7	113	9.2	595	26.7

The table 6.8 reveals about the persona hygienic practice of the respondents', 93.9 per cent of brushing their teeth once a day, and 10.9 per cent brushing their teeth twice a day. 32.1 per cent of cutting their nail weekly once, 14.8 per cent whenever necessary and whenever necessary, and 53.1 per cent of once in a month. 92.8 per cent of people hand wash whenever necessary, 3.3 per cent mentioned two hours once and every one hour once washed their hands. 34.0 per cent hair wash once a day, 9.3 per cent told twice a day, 22.5 per cent weekly once, and 34.2 per cent weekly twice washed their hair. 54.8 per cent of washed their clothes daily, 18.5 per cent washed twice a day and 26.7 per cent of washed once a day.

Literature emphasizes the importance of regular personal hygiene practices for health and well-being: Daily brushing reduces dental plaque and prevents oral diseases (Petersen & Ogawa, 2012). Regular hand washing with soap is crucial for preventing the spread of infections (Curtis & Cairncross, 2003). Frequency varies culturally and by climate, impacting personal comfort and cleanliness (Heymann, 2014).

Table 6.8 provides insights into various aspects of personal hygiene practices across three districts: Most respondents in Ariyalur (96.3%), Perambalur (89.2%), and Nilgiris (93.9%) brush their teeth once a day, with a smaller proportion brushing teeth twice a day, nail cutting practices vary, with a majority in each district opting to cut nails whenever necessary, the majority of respondents in all districts wash their hands whenever necessary, with varying frequencies of every one hour to two hours once, daily hair washing is predominant in Ariyalur (66.1%), while in Perambalur (58.2%) and Nilgiris (34.0%), weekly washing practices are more common, daily cloth washing is prevalent in Ariyalur (92.1%) and while in Perambalur (84.5%) and Nilgiris (54.8%), it is less frequent.

5.10 Menstrual Hygiene

Literature emphasizes cultural norms and societal perceptions influencing menstrual hygiene practices: Many cultures treat menstruation as a private matter or subject to social taboos (Van Eijk et al., 2013). Practices like menstrual restrictions can impact women's daily lives and health outcomes (Hennegan et al., 2019). Education and awareness programs can help challenge stigma and promote healthier menstrual practices (Sommer et al., 2015). Table 6.9 presents the attitudes and practices related to menstrual hygiene management (MHM) across three districts: A significant proportion of respondents in Ariyalur (43.7%), Perambalur (53.3%), and Nilgiris (43.9%) perceive MHM as a private matter. Many respondents in Ariyalur (37.3%), Perambalur (33.4%), and Nilgiris (28.8%) adhere to restrictions or practices related to menstruation. And a smaller percentage of respondents believe that restrictions related to menstruation are not necessary, with variations across districts (Ariyalur 26.5%, Perambalur 12.3%, Nilgiris 14.0%).

Table 6.9

District Wise percentage distribution of Menstrual Hygiene

Personal Hygiene	Ariyalur		Perambalur		Nilgiris	
	N	%	N	%	N	%
MHM is private issue	479	43.7	655	53.3	977	43.9
Observed restriction	409	37.3	411	33.4	642	28.8
Not necessary restriction	291	26.5	151	12.3	312	14.0

The table 6.3 multiple responses set reveals the suggestion for the sanitary napkin for use the menstrual period, 33.0 per cent suggest the eco-friendly recycle napkin, 22.0 per cent suggest any type of pad but not cause stain, 27.6 per cent suggest the physically healthy pads, 17.3 per cent suggest any pad which serves to handle the menstrual flow.

Table 6.10

District Wise percentage distribution of Suggestion of Sanitary Napkin

Personal Hygiene	Ariyalur		Perambalur		Nilgiris	
	N	%	N	%	N	%
Eco-friendly recyclable pads	460	88.1	349	50.0	766	79.8
Any pad- not cause stain	389	74.5	382	54.7	511	53.2
Physically healthy pads	431	82.6	366	52.4	641	66.8
Any pad which serves to handle the menstrual flow	317	60.7	419	60.0	401	41.8

Literature on menstrual hygiene emphasizes the importance of access to and preference for environmentally friendly and physically healthy menstrual products. Preferences often vary based on cultural, economic, and environmental factors, influencing choices in sanitary products.

Table 6.11

District Wise Multi Responses of Restriction of during menstruation

Restriction	Ariyalur		Perambalur		Nilgiris	
	N	%	N	%	N	%
No go outside	207	50.6	86	20.9	331	51.7
No visit to worship places	396	96.8	400	97.3	573	89.5
No touching of plants, children	250	61.1	124	30.2	364	56.9
Staying at home alone	158	38.6	81	19.7	153	23.9

Table 6.10 presents the district-wise percentage distribution of responses regarding suggestions of sanitary napkin for menstruation period Ariyalur, Perambalur, and Nilgiris districts. In Ariyalur, 88.1% of respondents prefer eco-friendly recyclable pads, compared to 50.0% in

Perambalur and 79.8% in Nilgiris suggest the eco-friendly recyclable pads. 74.5% of respondents in Ariyalur prefer any pad that does not cause stains, compared to 54.7% in Perambalur and 53.2% in Nilgiris suggest any-pad not cause stain, 82.6% of respondents in Ariyalur opt for physically healthy pads, while the percentages are 52.4% in Perambalur and 66.8% in Nilgiris physically healthy pads and 60.7% of respondents in Ariyalur prefer any pad that serves to handle menstrual flow, compared to 60.0% in Perambalur and 41.8% in Nilgiris suggest any pad which serves to handle the menstrual flow.

Literature on menstrual restrictions often discusses the cultural, religious, and social dimensions influencing women's behaviors during menstruation. Studies have shown that these practices vary widely across cultures and regions, influenced by beliefs about purity, religious norms, and social taboos (Sommer et al., 2015; Hennegan et al., 2019). Such practices can impact women's health, self-esteem, and opportunities for education and employment.

The table 6.11 presents data on the various restrictions observed by women during menstruation across three districts: Ariyalur, Perambalur, and Nilgiris. The responses are categorized into different restrictions, indicated by the number of respondents and the percentage within each district. In Ariyalur, 207 women (50.6%), in Perambalur 86 women (20.9%), and in Nilgiris 331 women (51.7%) reported restrictions on going outside during menstruation.

This suggests a significant cultural practice of limiting mobility during menstruation, particularly pronounced in Nilgiris. A large majority reported refraining from visiting worship places during menstruation, with 96.8% in Ariyalur, 97.3% in Perambalur, and 89.5% in Nilgiris, indicating strong cultural and religious beliefs influencing women's activities during this time. Restriction on touching plants or children was reported by 61.1% in Ariyalur, 30.2% in Perambalur, and 56.9% in Nilgiris, highlighting variations in adherence to this practice across districts. The practice of staying at home alone during menstruation was observed by 38.6% in Ariyalur, 19.7% in Perambalur, and 23.9% in Nilgiris, indicating differences in social norms regarding women's seclusion during menstruation.

Table 6.12

District Wise Muti Responses of the Suggestion for Toilet's Requirement in Educational Institutions and work place

Suggest of Educational Institutions/work place Toilets' requirement	Ariyalur		Perambalur		Nilgiris	
	N	%	N	%	N	%
Sanitary vending machine	427	89.1	512	78.0	694	73.0
Space for changing napkins	431	90.0	321	48.9	754	79.3
Adequate water supply	424	88.5	327	49.8	769	80.9
Facility to Disposed napkins	355	74.1	457	69.7	589	61.9
Periodical awareness needs	331	69.1	282	43.0	466	49.0
Separate Toilet building to women	267	55.7	260	39.6	380	40.0

The table 6.12 presents data on the suggestions for toilet requirements in educational institutions and workplaces across three districts: Ariyalur, Perambalur, and Nilgiris. The responses are categorized into various suggestions, with the number of respondents and the percentage within each district provided. A significant majority suggested the installation of sanitary vending machines, with 89.1% in Ariyalur, 78.0% in Perambalur, and 73.0% in Nilgiris. This indicates a strong demand for access to menstrual hygiene products in these settings. About 90.0% in Ariyalur, 48.9% in Perambalur, and 79.3% in Nilgiris emphasized the need for designated spaces for changing napkins, highlighting varying levels of awareness and infrastructure across districts. The requirement for adequate water supply was noted by 88.5% in Ariyalur, 49.8% in Perambalur, and 80.9% in Nilgiris, underscoring the essential need for reliable water access to support menstrual hygiene management. Approximately 74.1% in Ariyalur, 69.7% in Perambalur, and 61.9% in Nilgiris highlighted the necessity for proper facilities to dispose of used napkins, indicating gaps in sanitation infrastructure.

There was a call for periodic awareness programs, noted by 69.1% in Ariyalur, 43.0% in Perambalur, and 49.0% in Nilgiris, highlighting the importance of education and information dissemination on menstrual hygiene. The need for separate toilet facilities for women was mentioned by 55.7% in Ariyalur, 39.6% in Perambalur, and 40.0% in Nilgiris, suggesting a desire for privacy and gender-sensitive infrastructure. A similar study was also expressed their desire for additional features such as urinals, washing and bathing areas, and cloakrooms. At a few locations, residents also requested for small shops that sold menstrual products and other toiletries (IIHS, 2022).

5.11 Suggestion of improve the sanitation through MHM:

Literature highlights that comprehensive MHM programs, including education, accessibility to products, and support from communities and institutions, are critical for improving menstrual health outcomes and overall well-being (Sommer et al., 2015; Phillips-Howard et al., 2016). Table 6.13 presents data on various suggestions related to menstrual hygiene management (MHM) needs across three districts: Ariyalur, Perambalur, and Nilgiris. The responses highlight the perceived requirements for improving MHM within the context of total sanitation efforts. Significant proportions in all districts (ranging from 45.9% to 83.6%) suggest the need for educational programs targeting men to enhance their understanding and support for menstrual hygiene. High percentages (ranging from 68.2% to 86.9%) indicate a strong call for supportive measures tailored to women and girls, emphasizing the need for comprehensive support systems. A majority in each district (ranging from 60.1% to 75.5%) advocates for improved accessibility to sanitary pads, essential for effective menstrual hygiene management. There is a notable demand (ranging from 30.2% to 64.2%) for incinerators to safely dispose of used sanitary products, highlighting environmental and hygiene concerns. A significant portion (ranging from 23.3% to 59.5%)

stresses the importance of increasing general awareness and understanding about menstruation. A call for men to actively support women during menstruation is noted (ranging from 17.5% to 29.1%), promoting gender inclusivity and support.

Table 6.13

Multi Responses of Suggestion of MHM needs to total Sanitation

Suggestion of MHM needs	Ariyalur		Perambalur		Nilgiris	
	N	%	N	%	N	%
Extension education to men	68	45.9	60	51.7	320	83.6
Support women and girls	101	68.2	53	45.7	333	86.9
Help Get the sanitary pads easily	89	60.1	49	42.2	289	75.5
Provide incinerators	95	64.2	35	30.2	218	56.9
Understand about menstrual	88	59.5	27	23.3	108	28.2
Men, should help when menses	43	29.1	27	23.3	67	17.5
Provide scientific information on	22	14.9	49	42.2	93	24.3
Parents/teachers take response	18	12.2	18	15.5	118	30.8

A need for scientifically accurate information on menstrual hygiene is indicated (ranging from 14.9% to 42.2%), suggesting gaps in knowledge dissemination. A substantial percentage (ranging from 12.2% to 30.8%) suggests that parents and teachers should play a more active role in educating and supporting young girls regarding menstrual hygiene.

Findings and Conclusion

Ariyalur predominantly consists of Municipalities (68.8%), while Perambalur has a slight majority of Town Panchayats (52.5%). Nilgiris shows an equal split between Town Panchayats and Municipalities (50% each). Nilgiris has a higher male population (54.8%), Perambalur has more females (56.7%), and Ariyalur shows a balanced gender distribution with a slight male majority (52.3%). The majority of the population in all districts falls within the 25-54 age group, with Nilgiris having a higher proportion of elderly (65+) at 16.2% compared to Perambalur (15.2%) and Ariyalur (12.9%). Nuclear families are predominant across all districts, with Nilgiris and Ariyalur having the highest percentages (93.5% and 91.2%, respectively). Perambalur has a higher proportion of joint families (19.0%). Ariyalur and Perambalur have a Hindu majority, while Nilgiris has a diverse composition with a notable Christian population. Perambalur also has a higher percentage of Muslims.

Backward Classes (BC) dominate all three districts, with Scheduled Castes (SC) forming a significant portion in Nilgiris. Ariyalur shows a prevalence of skilled manual labor, while unskilled manual labor is more common in Perambalur and Nilgiris. Housewives constitute a notable demographic in Ariyalur. A substantial proportion of the population in all districts falls within the lowest income bracket (<=1000). Nilgiris shows a significant number in the middle-income range (5001-10000), indicating better economic conditions compared to Ariyalur and Perambalur. Concrete houses are predominant in all districts, indicating relatively good living conditions. However, Nilgiris has a notable proportion of asbestos/steel houses (22.3%). Homeownership is high across all districts, with Ariyalur leading at 81.3%. Rented houses are more common in Perambalur and Nilgiris. Almost all households in Ariyalur own their homestead land (99.8%), while Perambalur and Nilgiris rely more on government-owned land.

These findings underscore the diverse demographic and socio-economic landscapes across Ariyalur, Perambalur, and Nilgiris districts, highlighting areas for targeted interventions in infrastructure development, economic upliftment, and social welfare to enhance overall well-being and equity among residents.

Public taps are the dominant source for drinking and non-drinking purposes in Nilgiris (68.7%), whereas Ariyalur relies more on surface water sources like ponds and streams (90.9%). Perambalur shows a balanced reliance on public taps (18.5%) and borewells (54.3%). Individual household taps are prevalent in all districts, with Perambalur having the highest usage (35.4%) compared to Ariyalur (32.8%) and Nilgiris (31.8%). Ariyalur significantly uses surface water (90.9%), while Perambalur (64.7%) and Nilgiris (30.1%) show varied preferences, influenced by local geography and infrastructure.

Ariyalur and Perambalur experience limited daily water supply (1-5 hours/day) more frequently than Nilgiris, which generally enjoys better daily access. Weekly water supply irregularities are more pronounced in Ariyalur and Perambalur, impacting daily routines and health outcomes. Females predominantly handle water fetching responsibilities, particularly in Ariyalur and Perambalur, highlighting gender disparities in water management.

Boiling is the most common method across all districts (38.9%), followed by cloth or strainer filters (22.9%) and reverse osmosis (9.5%). Significant proportions (23.5%) do not treat their water, posing potential health risks. Nilgiris exhibits higher boiling practices (54.3%), while Ariyalur reports the highest non-treatment rate (47.6%) and lower adoption of reverse osmosis systems (3.6%). Colorless and transparent water is predominant (58.6%), indicating good quality. However, sediment (8.7%), colored water (8.4%), and unacceptable taste/odor (22.0%) are reported issues. Ariyalur shows higher sediment presence (14.6%) and better taste/odor perception (50.4%). Perambalur reports more colored (6.6%) and salty water (7.8%), while Nilgiris has the highest percentage of colorless water (75.1%) and the lowest chlorinated water presence (0.7%). These findings underscore the varied water source preferences, access challenges, treatmentpractices, and quality disparities across Ariyalur, Perambalur, and Nilgiris districts. Addressing these issues requires tailored interventions focusing on infrastructure development, equitable access to safe water, and community education on effective water management and treatment practices.

The data reveals high availability of toilets across all districts, with 83.0% of households in Nilgiris having access to toilets at home. Indian-style toilets are predominant, reflecting regional preferences and economic factors. The involvement of individuals in toilet construction varies, with notable female participation in Perambalur and Nilgiris. This underscores the importance of gender-sensitive approaches in sanitation initiatives. Most toilets were constructed within the past decade, indicating ongoing sanitation infrastructure development. Septic tanks are widely used, though maintenance practices vary. Significant percentages in all districts never clean their tanks, highlighting potential hygiene risks and the need for improved maintenance education and services.

While most respondents reported no major issues with their toilets, a minority faced challenges such as septic tank overflow, broken basins, and pipe blockages. Timely maintenance and education on proper usage and upkeep are crucial to mitigate these issues and ensure sustainable sanitation practices. Despite progress, open defecation remains a concern, particularly in Ariyalur (19.4%), Perambalur (12.5%), and Nilgiris (12.3%) districts. Locations such as bushes, empty plots, and roadside areas are commonly used, posing health risks and environmental challenges. Addressing this requires comprehensive strategies including community engagement, infrastructure improvements, and behavior change campaigns.

This chapter delves into various aspects of personal hygiene including food preparation, hand washing habits, bathing frequencies, and menstrual hygiene practices across three districts: Ariyalur, Perambalur, and Nilgiris. Hygienic practices during food handling are widely practiced, with a high percentage (over 95%) of respondents across districts covering cooked food adequately to prevent contamination. However, there are notable variations in other practices such as wearing protective gear and avoiding cooking while sick, suggesting areas for targeted education and reinforcement. Hand washing practices are generally well-adhered to, particularly before eating and after using the toilet, although there is room for improvement in consistent hand hygiene across different daily activities.

Personal hygiene practices like bathing and teeth brushing show cultural and regional variations. Daily bathing is predominant, especially in Ariyalur and Perambalur, while less frequent in Nilgiris where weekly bathing is more common. Access to private bathing facilities is widespread, although some households still rely on open spaces for bathing, indicating disparities in sanitation infrastructure. During menstruation, bathing habits vary, reflecting cultural norms and personal preferences, with some women maintaining regular bathing routines while others adhere to specific practices during their menstrual period.

Menstrual hygiene management (MHM) practices reveal significant cultural influences and preferences for sanitary products that are environmentally friendly and promote health. There is a strong call for improved facilities in educational institutions and workplaces, including access to sanitary vending machines, designated changing spaces, and adequate water supply. Awareness programs are also recommended to address misconceptions and promote better menstrual health practices.

The demographic and socioeconomic analysis of Ariyalur, Perambalur, and Nilgiris districts provides valuable insights into the residential patterns, demographic characteristics, and socioeconomic status of the populations within these regions. The distribution of residential areas varies significantly across the districts, with Nilgiris showing an equal split between Town Panchayat and Municipality areas, while Perambalur leans towards Town Panchayats and Ariyalur towards Municipalities. Demographically, gender distribution varies, with Nilgiris having a higher male population, Perambalur more females, and Ariyalur relatively balanced. The majority of households across all districts are nuclear families, with variations in household sizes and educational attainment levels.

Socioeconomically, religious composition, occupational distributions, and income brackets highlight distinct characteristics of each district. Ariyalur and Perambalur have a Hindu majority, whereas Nilgiris exhibits religious diversity with a notable Christian population. The dominance of manual labor in Ariyalur contrasts with the prevalence of skilled professions in Nilgiris. Income distribution shows disparities, with Nilgiris having a higher proportion in the middle-income bracket compared to Ariyalur and Perambalur.

Access to safe drinking water is essential for public health and well-being, yet significant disparities in water sources and quality persist across Ariyalur, Perambalur, and Nilgiris districts. The reliance on public taps varies widely, with Nilgiris showing substantial dependence, potentially indicating robust infrastructure or limited alternatives. In contrast, Perambalur's preference for borewells and surface water highlights differing availability and accessibility challenges. Socio-economic factors, including caste, play a crucial role in determining water access, with marginalized communities often facing greater barriers to safe water sources.

Water treatment practices also vary significantly, with boiling being the most common method across all districts. However, a concerning proportion of households do not treat their water, exposing them to potential health risks associated with contaminants. The presence of sediment, colored water, and unacceptable taste or odor underscores ongoing challenges in maintaining water quality standards, particularly in Perambalur and Ariyalur.

Efforts to improve water infrastructure, promote sustainable water management practices, and enhance community awareness about water treatment and hygiene are crucial. Targeted interventions should prioritize equitable access to safe drinking water, especially for marginalized groups, and strengthen water quality monitoring and treatment initiatives. By addressing these challenges comprehensively, we can work towards ensuring universal access to clean and safe drinking water across all districts, thereby improving health outcomes and enhancing overall quality of life.

The study provides a comprehensive assessment of sanitation practices and facilities across Ariyalur, Perambalur, and Nilgiris districts, highlighting both achievements and areas needing improvement. Access to toilets has significantly improved, with a high percentage of households having sanitation facilities, predominantly Indian-style toilets. However, challenges such as gender disparities in sanitation responsibilities and infrastructure maintenance issues persist. The data underscores the importance of sustainable sanitation management, including proper waste disposal through septic tanks and enhanced community education on hygiene practices. Addressing infrastructure defects in public and household toilets is critical to improving sanitation standards and reducing open defecation rates.

The study reveals significant insights into the environmental hygiene practices related to solid and liquid waste management across the Ariyalur, Perambalur, and Nilgiris districts. Awareness levels regarding solid waste management are relatively high, but disparities exist in waste segregation practices and disposal methods. While most households have access to waste collection services, improper disposal methods such as open dumping and burning still persist, indicating a need for targeted interventions to improve compliance with waste management regulations. Liquid waste management practices vary widely, with notable instances of unsafe disposal methods like pouring into open spaces and water bodies, posing risks to environmental and public health.

The findings from the study on personal hygiene practices across Ariyalur, Perambalur, and Nilgiris districts highlight both commendable adherence to certain hygienic behaviors and areas for improvement. Hygiene practices during food preparation, such as covering cooked food and maintaining clean hands, show generally high compliance but also reveal instances where education and reinforcement are needed. Hand washing practices, crucial for preventing disease transmission, are widely adopted at critical times but vary across specific activities like handling water or changing sanitary napkins, indicating opportunities for targeted hygiene education. Personal hygiene practices like bathing frequency and use of sanitation facilities reflect cultural norms and access disparities, with implications for health and well-being. Menstrual hygiene practices vary significantly, influenced by cultural beliefs and access to resources, suggesting a need for tailored education and infrastructure improvements to support women's health during menstruation.

Suggestions:

Since the creation of this earth on which we live, the five elements, which are the source of living beings, have renewed themselves and acquired a lasting character. This is nature's best sustainable scientific method. Unfortunate man failed to understand the nature of the earth with his knowledge and called what appeared to him as scientific knowledge and created a change in the land, water, air, and fire that existed since the beginning of the earth and spoiled its balance. When human science started in the world, nature started to perish. Nature has created a perfect scientific method and it has lasted for ages. But not only does man's knowledge fail to understand natural science, he destroys nature and seeks harm for himself without realizing that creating with his knowledge is an act of destroying nature. Land, air, water, fire and sky those five elements is base to the living things. Five elements are the source of living being. But Man pollutes the land, air, air and water with his knowledge and meets its effects. That is, he hurts his eyes with his own hand. Since both water and land are visible to the eye, man owns them as much as he wants with his strength and desire moreover human sense changed water and land nature and spoiled it. The result is that man has become unable to live by using them. So, this study suggested:

By selecting people who have western toilets and conducting further research on them, we can find out whether there is any relationship between Weston toilet effects, constipation problems, physical problems, maintenance challenges etc. This can lead to health problems and should be investigated. A large percentage mentioned that the septic tank is not always cleaned. This means that if it is not cleaned regularly, what becomes sewage, the septic tank water mixes with the ground water and would create a sanitary problem. It is necessary to know why people who do not want to build a toilet and make alternative arrangements, people who do not want to defecate in the open, this also needs to be investigated and action taken.

Prioritize infrastructure development in Perambalur to support the higher proportion of joint families and ensure adequate housing and amenities. Implement educational initiatives in Ariyalur to uplift the skill levels of the predominantly manual labor force, fostering economic growth and reducing dependency on unskilled jobs. Address housing concerns in Nilgiris where a significant portion of houses are made of asbestos/steel, by promoting safer housing alternatives and improving living conditions. Encourage economic diversification in all districts, focusing on sectors that can absorb different educational and skill backgrounds to improve income distribution and overall economic stability. Expand social welfare programs targeting vulnerable populations, especially in areas with higher proportions of SC and backward communities, to enhance social inclusion and equity. Foster community engagement and participation in decision-making processes related to infrastructure, health, and education to ensure local needs are addressed effectively. Implement awareness campaigns to encourage regular maintenance of toilets, including cleaning and septic tank management. Collaborate with local governments and NGOs to provide training on basic maintenance practices.

Address infrastructure defects in public toilets such as broken doors, insufficient water supply, and inadequate sanitation facilities. Prioritize the installation of essential amenities like hand dryers and soap dispensers to enhance user experience. Launch behavior changes campaigns targeting communities practicing open defecation. Focus on cultural norms, privacy concerns, and health impacts to encourage toilet adoption. Promote gender equality in sanitation practices by involving both men and women in decision-making processes and infrastructure development. Ensure that sanitation facilities meet the needs of all household members. Strengthen government initiatives and subsidies to facilitate toilet construction, particularly in underserved and rural areas. Ensure accessibility and affordability of sanitation facilities through financial assistance programs. Establish monitoring mechanisms to track progress in sanitation infrastructure development and usage. Conduct periodic evaluations to assess the effectiveness of interventions and adjust strategies as needed. Launch targeted awareness programs to educate communities about the importance of proper waste segregation and disposal practices. Utilize local media, community events, and educational institutions to disseminate information.

Invest in improving waste collection and disposal infrastructure, particularly in areas with inadequate services. Ensure regular maintenance of existing facilities to prevent breakdowns and service interruptions. Foster community engagement in waste management through initiatives such as volunteer clean-up drives, neighborhood composting programs, and involvement in local waste management committees. Strengthen enforcement of existing waste management regulations to discourage improper disposal practices. Implement penalties for violations and provide incentives for compliance with environmentally friendly practices. Support research initiatives to develop innovative waste management technologies and sustainable practices tailored to local needs and conditions. Launch targeted campaigns on proper food handling, hand washing, and menstrual hygiene practices tailored to local cultural norms.

Invest in sanitation infrastructure to ensure universal access to private bathing facilities and improve facilities in educational and workplace settings for better MHM. Involve community leaders and educators in promoting hygiene practices, especially focusing on changing behaviors related to food handling and menstrual hygiene. Launch comprehensive hygiene education programs focusing on proper hand washing techniques, hygiene during food preparation, and menstrual hygiene management.

Target schools, workplaces, and community centers to reach diverse populations. Invest in sanitation infrastructure, including accessible toilets, sanitary napkin disposal facilities, and water supply systems. Prioritize locations like schools and workplaces to ensure adequate facilities for menstrual hygiene management. Tailor educational materials and campaigns to respect local cultural practices and beliefs regarding hygiene, particularly during menstruation. Engage community leaders and influencers to promote acceptance and understanding of healthy hygiene practices. Ensure consistent availability of affordable and environmentally friendly menstrual hygiene products. Establish vending machines and distribution points in schools and public spaces to enhance accessibility. Implement monitoring mechanisms to track the adoption of recommended hygiene practices and infrastructure improvements.

Conduct periodic evaluations to assess the impact on public health outcomes and adjust strategies as needed. Implement regular inspection and maintenance schedules for toilets and septic tanks to prevent issues like overflow and blockages. Launch community-wide awareness programs emphasizing the health benefits of using toilets and the dangers of open defecation. Ensure sanitation programs consider gender dynamics, promoting equal participation in decision-making and infrastructure use. Collaborate with local communities, NGOs, and government bodies to foster ownership of sanitation initiatives and promote sustainable practices. Invest in improving sanitation infrastructure, including public toilets with enhanced accessibility features for vulnerable groups. Advocate for policies that incentivize toilet construction and maintenance, including subsidies and financing options for low-income households. Launch targeted awareness programs to educate communities about proper waste segregation, disposal methods, and the importance of environmental hygiene. Invest in improving waste collection systems, especially in areas with lower service coverage, to ensure regular and efficient waste removal. Encourage composting and the use of organic waste for manure to reduce landfill volumes and enhance soil fertility. Strictly enforce waste management regulations to curb improper disposal practices like open dumping and burning, ensuring compliance through penalties and incentives.

Foster community participation in waste management initiatives through partnerships with local authorities and educational institutions, empowering residents to take ownership of their environmental responsibilities.

The major conclusion drawn from this research stress the need for some of the following recommendations:

* People drink the water as it is without any purification. They also need advice and instructions, so it is necessary to hold awareness meetings for the people about water safety and drinking quality water.

* Should develop the rain water harvesting in Municipality administration level, and motivate the individual also.

* Should prevent the waste water generation.

* Should organize the awareness camp to the entire people get surface water.

* Ooty Municipality, Town Panchyat ward members should take effort to clean the surface water bodies and ratification the access and availability of drinking and non-drinking water.

* Repair of broken pipes, replacement of plastic pipes with copper pipes, more flexible.

* Construction of drinking water storage tanks and appropriate measures should be taken to prevent rodents from living in drinking water tanks.

* Ooty Municipality should increase water supply to meet the growing population.

* Refusal of entry to new immigrants to Ooty.

Reference:

* Ankitha Gattupalli (2023), Urban Sanitation in India: Past, Present, and Future, https://www.archdaily.com/

* Aiello, A. E., Coulborn, R. M., & Perez, V. (2008). Effect of hand hygiene on infectious disease risk in the community setting: A meta-analysis. American Journal of Public Health, 98(8), 1372-1381.

* Arunava Saha et.al. (2021), "Water quality, sanitation, and hygiene among the tribal community residing in Jawadhi hills,

Tamilnadu: An observational study from Southern India", Journal of Family Medicine and Primary Care, https://www.researchgate.net/publication/347784639,DOI: 0.4103/jfmpc.jfmpc_1519_20

* Arul Pennadic Kamaliya S, S.Elamathi, M.Kaviya, K.Keerthana, Mr.E.Suresh kumar (2017), "Water Scarcity Analysis for Perambalur Town by using Arc-Gis" SSRG International Journal of Geo informatics and Geological Science - (IJGGS).

* Bhusan vidhya and sachedeva D.R. (1984), "An Introduction to sociology", Allahabad, Kitab Mahal, P.565.

* Curtis, V., & Cairncross, S. (2003). Effect of washing hands with soap on diarrhoea risk in the community: A systematic review. The Lancet Infectious Diseases, 3(5), 275-281.

* Deborah Balk, Mark R. Montgomery, Hasim Engin, Natalie Lin, Elizabeth Major, Bryan Jones, 2019, "Urbanization in India: Population and Urban Classification Grids for 2011" Data Descriptor, MDPI,

* Dr.R.Elangovan 2021, "Water Water", Coimbatore Research Centre, Udumalaipettai

* D.Freeda Grana Rani, K.Arunkumar, T.Valarmathy, (2006), Potability of drinking water sources of eleven villages in Perambalur district, Tamil Nadu, Researchgate

* Fernanda Deister Moreira, Sonaly Rezende, Fabiana Passos, (2021), Public toilets from the perspective of users: a case study in a public place, Brazil, Journal Water Health, Jan; 20(1):41-53, https://pubmed.ncbi.nlm.nih.gov/35100153.

* Hennegan, J., Shannon, A. K., Rubli, J., Schwab, K. J., & Melendez-Torres, G. J. (2019). Women's and girls' experiences of menstruation in low- and middle-income countries: A systematic review and qualitative metasynthesis. PLOS Medicine, 16(5), e1002803.

* Heymann, D. L. (Ed.). (2014). Control of Communicable Diseases Manual. American Public Health Association.

* Hill, D., Farmer, T., & Matthews, R. (2007). Review of soaps for handwashing: Environmental aspects of hygiene and sanitation. International Journal of Environmental Health Research, 17(5), 319-340.

* Hoang Minh Giang, 2017, A Study on Development Methodology of Sustainable Solid Waste Management System by Using Multi-Objective DecisionMaking Model – A case study in Hoi An City, Vietnam, https://core.ac.uk/download/pdf/141513943.pdf)

* Kaushik Deka (2021), India's urban water crisis: Cities at risk of running out of water, Indiatoday Magazhine, New Delhi March.

* Muthachitra & B.Laxmi Kanta (2017), "An Analysis on Supply of Water in Tamil Nadu" Shanlax International Journal of Economics, Vol.5: No.3

* National Family Health Survey – 2015-2016, Ministry of Health and Family, Government of India

* National classification of occupation-2015, Vol – I, Government of India Ministry of Labour & Employment Directorate General of Employment www.ncs.go.in

* Navin Bhatt et.al., (2019), what motivates open defecation? A qualitative study from a rural setting in Nepal", https://doi.org/10.1371/journal.pone.0219246.

* Nallathambi Yoganath & Tarun Bhatnagar, (2018) Prevalence of open defecation among households with toilets and associated factors in rural south India: an analytical cross-sectional study, https://pubmed.ncbi.nlm.nih.gov/30032253/

* Nancy S, Suguna Elayaperumal, Amol R Dongre, (2022) Promotion of toilet construction and usage in rural Tamil Nadu: A mixed-methods evaluation study, https://pubmed.ncbi.nlm.nih.gov/37039168/

* Nilgiris District Diagnostic Studies-DDS, (2018) Vazhndhu Kattuvom Project, https://tamil.vkp-tnrtp.org

* Nilgiris Adivasi Welfare Association-NAWA, (March 10, 2021) Nilgiris tribal village becomes model for clean drinking water, The Times of India. https:// https://timesofindia.indiatimes.com/city/chennai/nilgiris-tribal-village)

* Nur Shahirah Abd Rahim & Norzila Othman, (2019) Home Water Purification System in Malaysia: Qualitativeand Quantitative Study, Materials Science and Engineering, https://iopscience.iop.org

* Petersen, P. E., & Ogawa, H. (2012). Prevention of dental caries through the use of fluoride – the WHO approach. Community Dentistry and Oral Epidemiology, 40(Suppl 2), 3-8.

* Phillips-Howard, P. A., Caruso, B., Torondel, B., Zulaika, G., Sahin, M., & Sommer, M. (2016). Menstrual hygiene management among adolescent schoolgirls in low- and middle-income countries: research priorities. Global Health Action, 9(1), 33032. doi:10.3402/gha.v9.33032.

* Raman Sivakumar, Rangaswamy Mohanraj and Azeez PA (2000) "Physico-chemical analysis of water sources of Ooty, South India, Environmedia Poll Res.19 (1): 143-146)

* Raja Sekaran N, (2019), Availability and Utilization of Sanitation Facilities: A Micro Study from Rural Tamil Nadu, International Journal of Economics, http://www.shanlaxjournals.in

* Ruth Maxine Flores, Vince Joell Feratero, Shant Kenneth C.Soneja, Dancia Bagus (2017), A Case Study about the Improper Waste Disposal in Barangay Mojon Tampoy, https://www.researchgate.net/publication/331702185

* Sanjay Singh, R., Sireesha, S., Premalatha, M., Gouru Naidu+, K and K.V. Ramesh1, "A study on drinking water quality at public

utilities in visakhapatnam, india", Global J. of Engg. & Appl. Sciences, 2015: 5 (1)

* Senthilnathan S & Azeez PA (1999), Water quality of effluents from dying and bleaching industry in Tirupur, Tamilnadu, India, Journal of Industrial Pollution Control, 15 (1), pp 79-88).

* Sommer, M., Caruso, B. A., Sahin, M., Calderon, T., Cavill, S., & Mahon, T. (2015). A Time for Global Action: Addressing Girls' Menstrual Hygiene Management Needs in Schools. PLoS Medicine, 12(6), e1001842. doi:10.1371/journal.pmed.1001842.

* Subramani, (2012) Study of pollution Preventions strategies for reclamation of Ooty Lake, International Journal of Engineering Research and Applications, Vol. 2, Issue 3, May-Jun 2012, pp.783-791, www.ijera.com

* Tamilarasan Muniyapillai, et.al. (2022), Water, sanitation and Hygiene (WASH) practices among Households in Perambalur District: A cross-sectional study", National Library of Medicine, National center for Biotechnology Information, http://www.ncbi.nlm.nih.gov/pmc/articles/PMC9643017

* United states Environmental Protection Agency-EPA (2023), Septic system Impacts on Water Sources, https://www.epa.gov.

* UNICEF (2018), Regional Office for South Asia. UNICEF South Asia / 2018–2021 PROGRESS REPORT [Internet]. Stop Open Defecation. http://www.unicefrosa-progressreport.org/opendefecation.html

* Yasemin Afacan & Meltem O Gurel (2015), Public toilets: an exploratory study on the demands, needs, and expectations in Turkey, Environment and Planning B: Planning and Design 2015, volume 42, pages 242–262, http://yoksis.bilkent.edu.tr/pdf/files/7424.pdf.

* Van Eijk, A. M., Sivakami, M., Thakkar, M. B., Bauman, A., Laserson, K. F., Coates, S., & Phillips-Howard, P. A. (2013).

Menstrual hygiene management among adolescent girls in India: A systematic review and meta-analysis. BMJ Open, 3(7), e002842.

Appendix

Drainage in Ooty Municipality

Water leakage (always) from the water Tank in Kethi Town Panchayat

Stream Water collecting Tank in Ooty

Ariyalur Municipality Lake

Public Toilet in Ooty Municipality

Public Toilet in Ariyalur Municipality

Public Toilet in Perambalur Municipality

Stream Water collecting Tank in Ooty

Drainage in Ooty Municipality

Mettur Middle School, Arumbavur Town Panchayat

Perambalur Municipality

Wastage dumping in Street

Wastage dumped in River in Perambalur

Wastage burned incuding plastics

Wastage dumping in drainage in Permbalur Municipality

******

FSC
www.fsc.org
MIX
Papier | Fördert
gute Waldnutzung
FSC® C083411

Kolibri 360 GmbH
Ferdinand-Jühlke-Straße 7
99095 Erfurt, Deutschland
produktsicherheit@kolibri360.de

Druck:
CPI Druckdienstleistungen GmbH
im Auftrag der
Kolibri 360 GmbH
Ferdinand-Jühlke-Str. 7
99095 Erfurt